Reviews of
121 Physiology Biochemistry and Pharmacology

Editors

M.P. Blaustein, Baltimore · H. Grunicke, Innsbruck
E. Habermann, Gießen · D. Pette, Konstanz
H. Reuter, Bern · B. Sakmann, Heidelberg
M. Schweiger, Innsbruck · E.M. Wright, Los Angeles

With 22 Figures and 3 Tables

Springer-Verlag Berlin Heidelberg GmbH

ISBN 978-3-662-31150-9   ISBN 978-3-540-47334-3 (eBook)
DOI 10.1007/978-3-540-47334-3

Library of Congress-Catalog-Card Number 74-3674

© Springer-Verlag Berlin Heidelberg 1992
Originally published by Springer-Verlag Berlin Heidelberg New York in 1992.
Softcover reprint of the hardcover 1st edition 1992
The use of registered names, trademarks, etc. in this publication does not imply, even in the
absence of a specific statement, that such names are exempt from the relevant protective laws
and regulations and therefore free for general use.

Product liability: The publishers cannot guarantee the accuracy of any information about
dosage and application contained in this book. In every individual case the user must check
such information by consulting the relevant literature.

Typesetting: Ulrich Kunkel Textservice, Reichartshausen, FRG
27/3130-5 4 3 2 1 0 – Printed on acid-free paper

Contents

Indexed in Current Contents

Rev. Physiol. Biochem. Pharmacol., Vol. 121
© Springer-Verlag 1992

Leukotrienes: Biosynthesis, Transport, Inactivation, and Analysis

DIETRICH KEPPLER

Contents

Division of Tumor Biochemistry, Deutsches Krebsforschungszentrum, Im Neuenheimer Feld 280, W-6900 Heidelberg 1, FRG

1 Introduction

The leukotrienes comprise a group of biologically highly potent mediators synthesized from 20-carbon polyunsaturated fatty acids, predominantly from arachidonate (Samuelsson et al. 1979; Murphy et al. 1979; Samuelsson 1983; Hammarström 1983). They include the cysteinyl leukotrienes LTC_4, LTD_4, LTE_4, and N-acetyl-LTE_4, as well as dihydroxyeicosatetraenoate leukotriene B_4 (LTB_4). Leukotrienes act at nanomolar concentrations in host defense, intercellular communication, and in signal transduction. The cysteinyl leukotrienes induce smooth muscle conctraction and increase vascular permeability (Dahlén et al. 1981; Lewis and Austen 1984; Piper 1984); LTB_4 elicits leukocyte sticking to vascular endothelia and inflammatory infiltration, and contributes in vivo to vascular permeability changes, immunoregulation, and pain responses (Ford-Hutchinson 1990, 1991b). Leukotrienes have been implicated as mediators in the pathogenesis of inflammatory, allergic, and other diseases, including bronchial asthma, arthritis, inflammatory bowel disease, anaphylaxis, shock, hepatorenal syndrome, pancreatitis, psorisasis, and tissue trauma (Piper 1984; Lewis and Austen 1984; Denzlinger et al. 1985; A. Keppler et al. 1987; Samuelsson et al. 1987; Keppler 1988, Huber et al. 1989; Huber and Keppler 1990; Ford-Hutchinson 1990). Only a limited number of cell types are capable of synthesizing LTC_4, LTB_4, or both. Predominant producer cells are macrophages, monocytes, neutrophils, eosinophils, mast cells, and basophils (Lewis and Austen 1984; Verhagen et al. 1984; Lewis et al. 1990). In addition, transcellular synthesis from the 5,6-epoxide LTA_4 released from some cells represents a pathway for synthesis of LTB_4 and LTC_4 in endothelial cells, platelets, mast cells, lymphocytes, and even erythrocytes (Odlander et al. 1988; Dahinden and Wirthmueller 1990; Feinmark 1990; Jones and Fitzpatrick 1990).

Recent progress in leukotriene research has led to a more detailed understanding of the enzymes and proteins mediating the biosynthesis of leukotrienes and to the development of potent inhibitors of biosynthesis as well as receptor antagonists interfering with signal transduction (for reviews see Rokach 1989; Piper and Krell 1991). Moreover, the mechanisms of leukotriene transport during release from biosynthetic cells (Lam et al. 1989, 1990; Schaub et al. 1991) and during hepatobiliary elimination (Ishikawa et al. 1990; Keppler et al. 1992) have been recognized, and pathways and compartmentation of leukotriene inactivation were further elucidated (Soberman et al. 1988; Stene and Murphy 1988; Keppler et al. 1989; Shirley and Murphy 1990; Sala et al. 1990; Jedlitschky et al. 1991). In addition to receptor-mediated leukotriene actions on the cell surface (Saussy et al.

1989; Herron et al. 1992), intracellular leukotriene actions in growth factor signal transduction have been recognized (Peppelenbosch et al. 1992).

2 Leukotriene Biosynthesis

Leukotriene biosynthesis is triggered under pathophysiological and experimental conditions by a variety of immunological and nonimmunological stimuli, including Ca^{2+} ionophores. The key enzyme, arachidonate 5-lipoxygenase (EC 1.13.11.34) depends on the availability of arachidonate, which is released from membrane phospholipids by phospholipase A_2 (EC 3.1.1.4) or by the sequential action of phospholipase C (EC 3.1.4.3) and diaclyglycerol lipase (EC 3.1.1.34). Ca^{2+}-dependent activation of phospholipase A_2 with subsequent release of arachidonate is associated with phosphorylation and translocation of the cytosolic phospholipase A_2 to membrane vesicles (Lin et al. 1992). The concentration of free arachidonate is controlled, in addition, by its reincorporation into lysophospholipids (Ferber and Resch 1973; Irvine 1982). Arachidonate 5-lipoxygenase is a

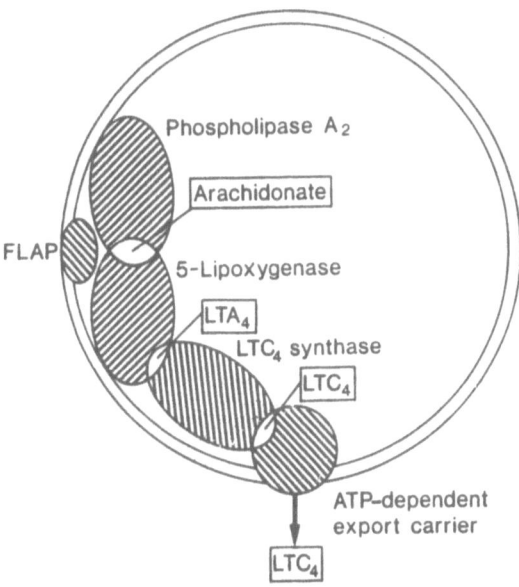

Fig. 1. Proposed scheme of the association and concerted action of enzymes and proteins involved in the synthesis of the parent cysteinyl leukotriene LTC_4. This association may allow for chanelling of the intermediates into the export carrier and may be localized in vesicles at the plasma membrane. FLAP designates the five-lipoxygenase-activating protein (Ford-Hutchinson 1991a). Release of LTC_4 from a leukotriene-synthesizing cell is mediated by an ATP-dependent export carrier (Schaub et al. 1991) which is distinct from the LTB_4 transporter

bifunctional enzyme and also catalyzes the synthesis of the 5,6-epoxide LTA_4. Depending on the differentiation of a leukotriene-generating cell, LTA_4 may be converted to LTB_4 by LTA_4 hydrolase (EC 3.3.2.6) or to LTC_4 by the membrane-bound enzyme LTC_4 synthase (EC 2.5.1.37). A protein termed *five-lipoxygenase-activating protein* (FLAP) is required, in addition, for the synthesis of LTC_4 and LTB_4 in intact cells (Dixon et al. 1990; Miller et al. 1990; Rouzer et al. 1990; Ford-Hutchinson 1991a). One may assume that the enzymes and proteins required for cellular synthesis of leukotrienes are closely associated and translocated to the cell membrane allowing for concerted catalysis and export from the cell (Fig. 1).

2.1 Biosynthetic Enzymes

Arachidonate 5-lipoxygenase catalyzes the first step in leukotriene synthesis by addition of oxygen to carbon 5 of arachidonate yielding (5*S*)-hydroperoxy-6,8,11,14-eicosatetraenoate. The latter is converted by the second catalytic activity of the 5-lipoxygenase protein, LTA_4 synthase, to 5,6-oxido-7,9,11,14-eicosatetraenoate (for review see Samuelsson and Funk 1989). 5-Lipoxygenase/LTA_4 synthase has been cloned and expressed in mammalian cells (Dixon et al. 1988; Matsumoto et al. 1988; Rouzer et al. 1988). This 78-kDa protein requires Ca^{2+} and ATP for maximal activity. Moreover, translocation of 5-lipoxygenase from the cytosol to the cell membrane, which is triggered by Ca^{2+}, is associated with activation of cellular leukotriene synthesis (Rouzer and Kargman 1988). Inhibition of this translocation by the indole derivative MK-886 inhibits leukotriene synthesis in intact cells (Rouzer et al. 1990). The target protein of MK-886 has been identified as the 18-kDa membrane protein FLAP, which is essential for leukotriene synthesis and must be coexpressed together with 5-lipoxygenase (Dixon et al. 1990; Miller et al. 1990; Reid et al. 1990; Ford-Hutchinson 1991a). FLAP may act to couple phospholipase A_2, membrane phospholipids, and 5-lipoxygenase. The presence of 5-lipoxygenase and FLAP is limited mostly to cells of the myeloid lineage and is related to cell differentiation (Habenicht et al. 1989).

The product of 5-lipoxygenase, LTA_4, is converted enzymatically either by LTA_4 hydrolase to LTB_4, by LTC_4 synthase to the glutathione conjugate LTC_4, by 15-lipoxygenation to 15-hydroxy-LTA_4, or by cytosolic epoxide hydrolase to 5(*S*),6(*R*)-dihydroxyeicosatetraenoate. *LTA_4 hydrolase* is a cytosolic monomeric protein of about 69 kDa which has been cloned and expressed in *Escherichia coli* (Samuelsson and Funk 1989). LTA_4 hydrolase has been detected in virtually all tissues as well as in blood plasma

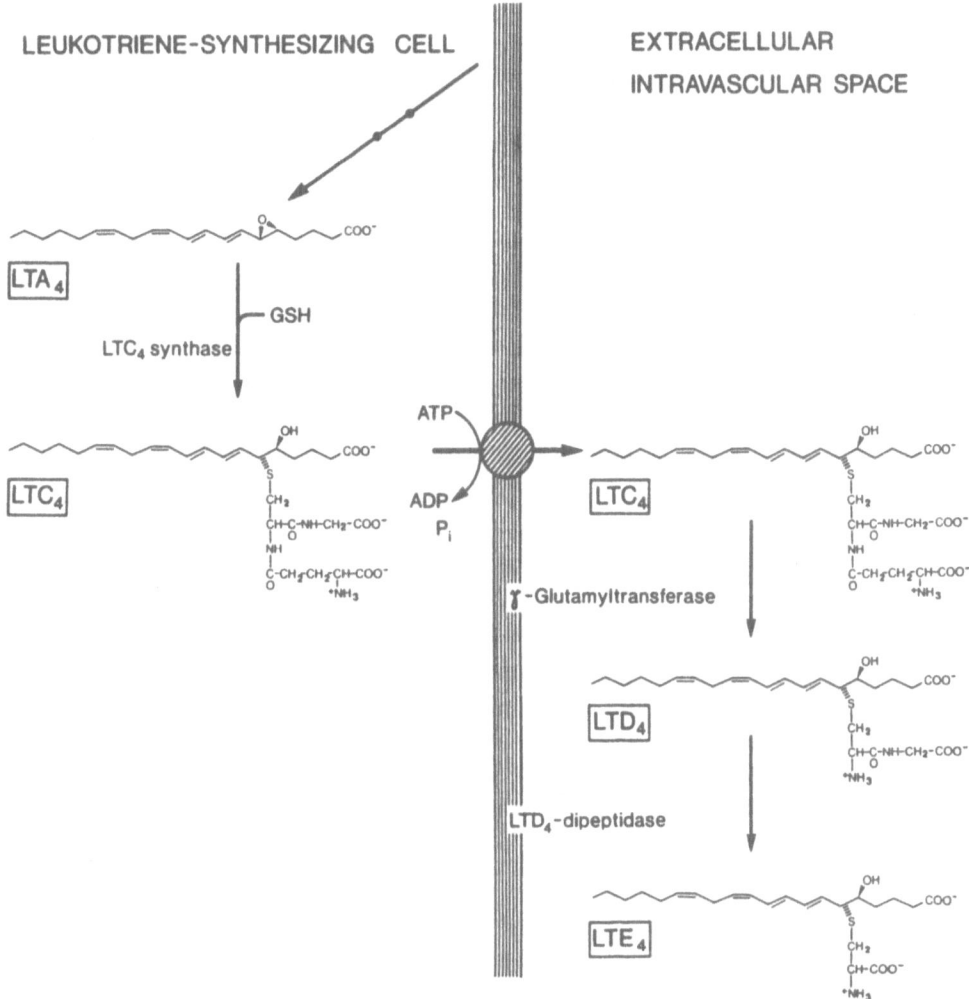

Fig. 2. Synthesis, export, and peptidolytic degradation of LTC_4. Synthesis of LTC_4 from LTA_4 and glutathione by microsomal LTC_4 synthase is followed by unidirectional ATP-dependent export from leukotriene-synthesizing cells, such as murine mastocytoma cells (Schaub et al. 1991). The ectoenzymes γ-glutamyltransferase and LTD_4 dipeptidase catalyze the biological activation and deactivation to LTD_4 and LTE_4, respectively (Hammarström et al. 1985)

from several species and in erythrocytes (McGee and Fitzpatrick 1985). Surprisingly, LTA_4 hydrolase has been identified as a Zn^{2+}-containing aminopeptidase with a sequence homologous to the active site of certain peptidases (Haeggström et al. 1990; Minami et al. 1990). Accordingly, the aminopeptidase inhibitor bestatin (Örning et al. 1991a) as well as the angiotensin-converting enzyme inhibitor captopril (Örning et al. 1991b) were found to act as inhibitors of LTB_4 synthesis from LTA_4 in the micromolar concentration range.

The synthesis of LTC_4 is catalyzed by membrane-bound *LTC_4 synthase* (Fig. 2), which is distinct from cytosolic and microsomal glutathione *S*-transferases (Söderström et al. 1988; Yoshimoto et al. 1988). The enzyme is highly specific for its substrate LTA_4 and has an isoelectric point of about 6, whereas other members of the glutathione *S*-transferase family are basic, with isoelectric points at or above 8.5, and catalyze the synthesis of a wide range of xenobiotic and endogenous glutathione *S*-conjugates (Söderström et al. 1988). LTC_4 synthase also reacts with LTA_3 and LTA_5. LTA_3 is a potent competitive inhibitor of LTC_4 synthesis from glutathione and LTA_4 (Yoshimoto et al. 1988). Further properties of LTC_4 synthase will be elucidated when this protein has been purified to homogeneity, cloned, and expressed. LTC_4 synthase is present not only in cells of the myeloid lineage, including mast cells and eosinophils, but also in several tissues and in endothelial cells (Feinmark 1990).

A *γ-glutamyltransferase* catalyzes the conversion of LTC_4 to LTD_4. Since LTD_4 is biologically much more potent than LTC_4 (Lewis and Austen 1984; Piper 1984), the partial degradation of the glutathione moiety to the cysteinylglycine derivative LTD_4 (Fig. 2) may be considered a bio-synthetic reaction generating the ligand for the LTD_4/LTE_4 receptor. γ-Glutamyltransferase is a glycoprotein enzyme widely distributed on cell surfaces. It has not been established whether a specific γ-glutamyltransfer-ase isoenzyme is responsible for LTD_4 generation. This reaction depends on catalysis in the low nanomolar concentration range by a high-affinity ectoenzyme (Weckbecker and Keppler 1986; Huber and Keppler 1987).

2.2 Transcellular Leukotriene Synthesis

Interaction between different cell types allows for enzymatic cooperation in leukotriene synthesis, also termed transcellular synthesis (Dahinden et al. 1985; McGee and Fitzpatrick 1986; Odlander et al. 1988; Dahinden and Wirthmueller 1990; Feinmark 1990; Jones and Fitzpatrick 1990). In neutrophils, LTA_4 formed in excess of the capacity for intracellular LTB_4 synthesis is released into the extracellular fluid where it can be stabilized by albumin. Neutrophil-derived LTA_4 is a precursor for leukotriene syn-thesis particularly in cell types deficient in 5-lipoxygenase, such as ery-throcytes, platelets, and vascular endothelial cells. As an example, LTA_4 hydrolase in erythrocytes generates LTB_4 from neutrophil-derived LTA_4 (McGee and Fitzpatrick 1986). Moreover, LTA_4, released from neutrophils and bound to albumin, serves in the synthesis of LTC_4 by mast cells

(Dahinden et al. 1985; Dahinden and Wirthmueller 1990). Thereby, the capacity of mast cells for LTC_4 generation is augmented. Leukotriene production under conditions where cell-cell cooperation occurs differs quantitatively and qualitatively from the sum of the separate cellular biosynthetic capacities. In disease processes different cell combinations may exist as compared to the normal condition. Transcellular leukotriene synthesis not only contributes to systemic leukotriene production but also influences the efficacy of inhibitors of leukotriene biosynthesis, which differ in their action on different cell types.

2.3 Inhibition of Leukotriene Biosynthesis and Action

Selective inhibition of leukotriene biosynthesis or selective blockade of the receptors for LTD_4 and LTE_4 or for LTB_4 is not only of therapeutic interest. These approaches furthermore serve to define the role of the leukotrienes under pathophysiological conditions. The recent development of biosynthesis inhibitors and of receptor antagonists has resulted in a considerable increase in selectivity and in compounds which are effective in the low nanomolar concentration range (Fitzsimmons and Rokach 1989; Ford-Hutchinson 1991a, b; Aharony and Krell 1991).

Direct inhibitors of 5-lipoxygenase have been described and many of them are "redox" inhibitors, presumably reducing the iron at the active site of the enzyme (for review see Fitzsimmons and Rokach 1989). Natural lipoxygenase inhibitors include the flavonoid compounds and hydroxylated cinnamic acids with cirsiliol and caffeic acid, respectively, as potent representatives in both groups. Nordihydroguaiaretic acid (NDGA) is a commercial antioxidant which inhibits lipoxygenase enzymes. NDGA is a widely used antioxidant inhibitor which lacks sufficient selectivity for 5-lipoxygenase and efficacy in the living mammalian organism. Among the quinone inhibitors Takeda's AA-861 represents a prototype compound with limitations comparable to NDGA. These compounds may be valuable in studies with cells in culture, but, in addition to a number of side effects, they do not sufficiently suppress systemic leukotriene production in the intact organism. Direct 5-lipoxygenase inhibitors that are effective in vivo and exhibit sufficient selectivity include compounds with hydroxamate or *N*-hydroxyurea functionalities. One of these drugs- A-64077 or zileuton [*N*-(1-benzo-thien-2-ylethyl)-*N*-hydroxyurea] is an effective inhibitor of leukotriene biosynthesis in man (Bell et al. 1992). As indicated in Fig. 3, this leads to an inhibition of both LTB_4 and systemic LTC_4 synthesis.

As an alternative approach to direct enzyme inhibition, *interference of 5-lipoxygenase translocation* to the plasma membrane by compounds

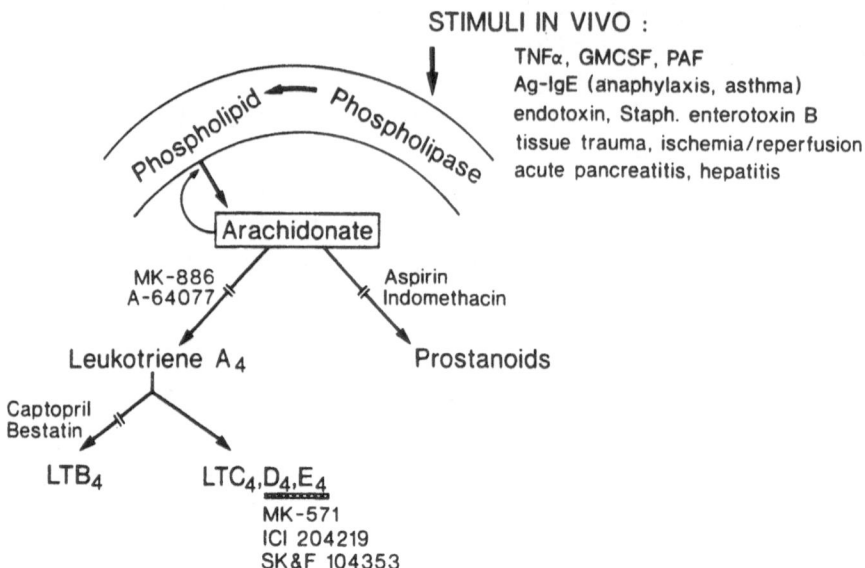

STIMULI IN VIVO :

TNFα, GMCSF, PAF
Ag-IgE (anaphylaxis, asthma)
endotoxin, Staph. enterotoxin B
tissue trauma, ischemia/reperfusion
acute pancreatitis, hepatitis

Fig. 3. The arachidonate cascade with several of its stimuli in the intact mammalian organism and with the sites of action of biosynthetic inhibitors and examples for LTD_4/LTE_4 receptor antagonists. Synthesis of leukotrienes may be elicited in vivo by a number of pathophysiological stimuli (for review see Keppler 1988) as well as physiological primers and eliciters such as the granulocyte-macrophage conoly-stimulating factor (*GMCSF*; Denzlinger et al. 1990). Inhibitors of leukotriene biosynthesis from the first generation of drug development include MK-886 (Gillard et al. 1989; Ford-Hutchinson 1991a) and A-64077 (Bell et al. 1992), as well as captopril and bestatin with their additional potential to inhibit LTA_4 hydrolase (Örning 1991a, b). More potent, second generation inhibitors of LTA_4 synthesis include MK-591, A-78773, and ICI's D-2138. Some selective and potent antagonists of LTD_4/LTE_4 receptors, MK-571, ICI 204219, and SK&F 104353, are indicated (Snyder and Fleisch 1989; Piper and Krell 1991; Lewis et al. 1991)

which bind to FLAP induces potent and selective suppression of the synthesis of leukotrienes (Gillard et al. 1989; Ford-Hutchinson 1991a; Evans et al. 1991). The indole derivative MK-886 (Gillard et al. 1989), which binds with high affinity to FLAP (Rouzer et al. 1990), does not significantly affect 5-lipoxygenase itself but blocks leukotriene synthesis in intact cells and in vivo. Systemic leukotriene production, measured by an index metabolite in bile during guinea pig anaphylaxis, is completely suppressed by MK-886 (Guhlmann et al. 1989). MK-886 also suppresses cysteinyl leukotriene excretion into human urine to a large extent (Ford-Hutchinson 1991a). FLAP, as a novel drug target for inhibiting the biosynthesis of leukotrienes, also binds a group of quinoline derivatives which inhibit leukotriene synthesis in intact cells with a similar mechanism of action and at lower concentrations than MK-886 (Evans et al. 1991).

Selective inhibition of LTB$_4$ biosynthesis, without inhibition of LTC$_4$ synthesis, has become feasible as a consequence of the discovery that bestatin and captopril inhibit LTA$_4$ hydrolase with IC$_{50}$ concentrations of 4 and 11 μM, respectively (Örning et al. 1991a, b). The functional resemblance of LTA$_4$ hydrolase to metallohydrolase enzymes (Haeggström et al. 1990) will necessitate chemical modification of the drugs for successful and selective inhibition of LTB$_4$ synthesis in vivo, without or with little inhibition of peptidases and angiotensin-converting enzyme.

Inhibition of LTB$_4$ action has been achieved by the development of LTB$_4$ receptor antagonists. Among these, the hydroxyacetophenone derivative LY 255283 has 50% inhibitory potency in the binding assay at a concentration of 87 nM (Herron et al. 1992).

LTD$_4$/LTE$_4$ receptor antagonists are, at present, most promising in drug development for antiasthma therapy (Piper and Krell 1991). Reasons to develop LTD$_4$/LTE$_4$ receptor antagonists have included the lack of evidence for signal transduction via LTC$_4$ receptors in man, the higher biological potency of LTD$_4$ relative to LTC$_4$, and the rapid formation of LTD$_4$ and LTE$_4$ from LTC$_4$ on cell surfaces and in the blood circulation in vivo. At least three new structural classes of high-affinity LTD$_4$/LTE$_4$ receptor antagonists have been developed and tested in man. These are SK&F 104353, ICI 204,219, and MK-571 (for reviews see Snyder and Fleisch 1989; Piper and Krell 1991; Lewis et al. 1991). These third-generation LTD$_4$/LTE$_4$ receptor antagonists are several orders of magnitude more potent and display a several hundred-fold improvement in their selectivity for LTD$_4$/LTE$_4$ receptors than the first antagonist, FPL 55712, developed in 1973 against slow-reacting substance of anaphylaxis (Augstein et al. 1973). Affinities of these antagonists were determined in the low nanomolar concentration range for the LTD$_4$/LTE$_4$ receptor in human airways and guinea pig trachea (Aharony and Krell 1991). Doses that are 50% effective in vivo after intravenous administration in the guinea pig are 46 nmol/kg for the indole-based ICI 204,219, 2 nmol/kg for the quinoline-based analog MK-571, and 550 nmol/kg for the LTD$_4$/LTE$_4$ analog SK&F 104353 (Aharony and Krell 1991). These compounds act as competitive antagonists, are highly effective in man, and contribute to a definition of LTD$_4$-mediated pathophysiological processes.

3 Transport of Leukotrienes During Biosynthesis and Excretion

Transport controls not only the release of LTC_4 (Lam et al. 1989; Schaub et al. 1991) and LTB_4 (Lam et al. 1990) from leukotriene-generating cells but also the removal of these mediators from the blood circulation in vivo. The liver is the most active organ for uptake, metabolic inactivation, and biliary excretion of leukotrienes (Appelgren and Hammarström 1982; Keppler et al. 1985; Hagmann et al. 1989; Wettstein et al. 1989). In addition, transport during renal excretion and during the limited intestinal reabsorption of cysteinyl leukotrienes contributes to the control of leukotriene concentrations in body fluids.

3.1 The Export Carrier Releasing LTC_4 After Its Biosynthesis

The release of LTC_4 has been studied in cultured human eosinophils incubated with exogenous LTA_4 (Lam et al. 1989). This transport is saturable, temperature-dependent, and inhibited by intracellular LTC_5, suggesting a carrier mediated process. The mechanism underlying the export of LTC_4 has been elucidated in plasma membrane vesicles prepared from murine mastocytoma cells and characterized as a primary-active, ATP-dependent process with apparent K_M values of 48 μM for ATP and 110 nM for LTC_4 (Schaub et al. 1991). Among the cysteinyl leukotrienes, LTC_4 is the best substrate for this ATP-dependent export carrier (Fig. 2). The relative transport rates at a concentration of 10 nM are 1.00, 0.31, 0.12, and 0.08 for LTC_4, LTD_4, LTE_4, and N-acetyl-LTE_4, respectively (Schaub et al. 1991). LTC_4 transport is competitively inhibited by the glutathione S-conjugate S-(2,4-dinitrophenyl)glutathione, and by several other amphiphilic anions including LTD_4/LTE_4 receptor antagonists (Schaub et al. 1991). Primary-active ATP-dependent transport is insignificant with LTB_4 as a substrate. Therefore, inhibition of the LTC_4 export carrier in leukotriene-synthesizing cells by structural analogs and LTD_4/LTE_4 receptor antagonists may serve as a novel pharmacological approach to interfere selectively with LTC_4 production without influencing LTB_4 generation. Isolation and molecular characterization of the LTC_4 export carrier from leukotriene-generating cells, such as mast cells, eosinophils, and monocytes, will answer the question whether this carrier belongs to the family of the ATP-dependent glutathione conjugate export carrier originally described in the erythrocyte plasma membrane (Kondo et al. 1980).

3.2 Leukotriene Uptake into Hepatocytes

Leukotrienes released into the blood circulation, with or without prior or subsequent interaction with leukotriene receptors, undergo rapid elimination from blood predominantly due to uptake by the liver (Appelgren and Hammarström 1982; Hagmann et al. 1984; Denzlinger et al. 1985; Huber and Keppler 1990; Hagmann and Korte 1990). Albumin serves as transport protein in the blood circulation (Falk et al. 1989). Uptake by hepatocytes has been demonstrated both for cysteinyl leukotrienes (Ormstad et al. 1982; Uehara et al. 1983; Weckbecker and Keppler 1986; Leier et al. 1992) and for LTB_4 (Hagmann and Korte 1990; Leier et al. 1992). Uptake of LTC_4, LTD_4, LTE_4, and N-acetyl-LTE_4 across the sinusoidal (basolateral) membrane into hepatocytes is independent of a Na^+-gradient and a K^+-diffusion potential (Leier et al. 1992). The uptake may be driven by high-affinity binding to intracellular proteins and by the unidirectional, ATP-dependent transport across the canalicular (apical) membrane into bile (Fig. 4; Ishikawa et al. 1990). At a concentration of 10 nM, the relative uptake rates into rat hepatocytes for LTC_4, LTD_4, LTE_4, and LTB_4 are 1.0, 1.3, 1.6 and 1.6, respectively. The K_M values for the leukotrienes range between 100 and 200 nM (Leier et al. 1992). Leukotriene-binding proteins possibly involved in hepatocellular transport were identified by the method of direct photoaffinity labeling in the deep-frozen state using the ^{3}H-labeled leukotriene itself as the photolabile ligand (Falk et al. 1989; Müller et al. 1991b, Leier et al. 1992). Liver membrane subfractions enriched with sinusoidal plasma membranes contain a 48-kDa polypeptide labeled both with [3H]LTE_4 and [3H]LTB_4. This polypeptide is not labeled by cysteinyl leukotrienes in hepatoma cells which are deficient in cysteinyl leukotriene uptake (Müller et al. 1991b; Leier et al. 1992). There is no convincing

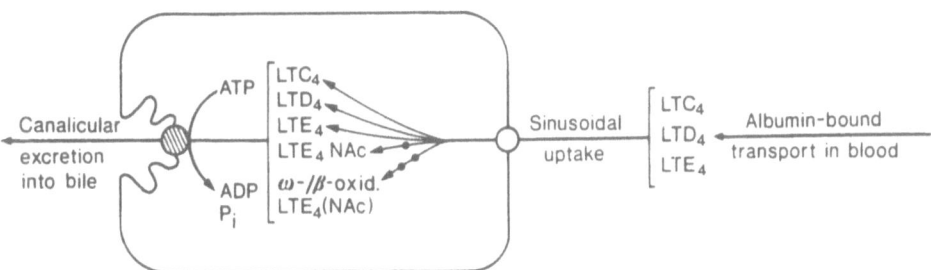

Fig. 4. Transport of cysteinyl leukotrienes · through hepatocytes. Uptake across the sinusoidal membrane may be followed by intracellular degradation (Keppler et al. 1989; Jedlitschky et al. 1991) and ATP-dependent export across the canalicular membrane into bile (Ishikawa et al. 1990). The latter process may be rate-limiting in overall hepatobiliary cysteinyl leukotriene elimination

evidence, however, that his 48-kDa polypeptide represents the transporter responsible for leukotriene uptake. It may rather be an intracellular, membrane-associated polypeptide binding the leukotrienes (Leier et al. 1992) as well as related amphiphilic substances (Kurz et al. 1989). The dihydroxy fatty acid LTB_4 differs from the cysteinyl leukotrienes by its entry into hepatoma cells, possibly as a result of its facilitated diffusion (Leier et al. 1992). Kinetic studies in hepatocytes employing inhibitors indicate the existence of distinct uptake systems for the cysteinyl leukotrienes and LTB_4 in the sinusoidal membrane. The substrate specificity of the transporters involved in leukotriene uptake across the sinusoidal hepatocyte membrane will be defined more precisely after reconstitution of the purified transporter in liposomes. The interaction of both cysteinyl leukotrienes and LTB_4 with hepatocytes does not lead to detectable receptor-mediated signal transduction if the mediators are added in the physiological nanomolar concentration range. This indicates that the hepatocyte uptake systems are transporters and not receptors for the leukotrienes.

3.3 The Cysteinyl Leukotriene Export Carrier in the Hepatocyte Canalicular Membrane

During the vectorial transport across the hepatocyte some of the leukotriene metabolites retain their structure and some undergo oxidative degradation from the ω-end (Figs. 4, 5). Products of ω- and β-oxidation of LTE_4, N-acetyl-LTE_4, and LTB_4, as well as unmodified LTC_4, LTD_4, LTE_4, and N-acetyl-LTE_4, are substrates for the leukotriene export carrier in the canalicular (apical) membrane of hepatocytes (Ishikawa et al. 1990). The mechanisms of this transport has been analyzed by use of plasma membrane vesicles enriched in canalicular membranes. The inside-out vesicles incubated in the presence of labeled cysteinyl leukotrienes and ATP showed primary-active, ATP-dependent uptake, corresponding to ATP-dependent export across the canalicular membrane into bile (Ishikawa et al. 1990). Primary-active, ATP-dependent transport seems to be domain-specific with a location in the canalicular but not in the sinusoidal hepatocyte membrane (Fig. 4). This is indicated by transport studies in vesicle preparations from different membrane domains (Ishikawa et al. 1990) and by photoaffinity labeling with the ^{35}S-labeled ATP analog ATP-γ-S of canalicular and sinusoidal membranes (Müller et al. 1991a). Among the cysteinyl leukotrienes, LTC_4 is the best substrate for the canalicular export carrier. Apparent K_M values are 0.25, 1.5, and 5.2 μM for LTC_4, LTD_4, and N-acetyl-LTE_4, respectively, whereas the K_M value for the cysteine S-conjugate, LTE_4, is more than 10 μM (Ishikawa et al. 1990). In addition, ω-carboxy-LTB_4, but not LTB_4 itself, is a substrate for ATP-dependent trans port across the canalicular membrane. Mutual competition among the cysteinyl leukotrienes and between leukotrienes and several glutathione S-conjugates and glucuronate conjugates suggests a common export carrier (Ishikawa et al. 1990; Akerboom et al. 1991). The term leukotriene export carrier is preferred since LTC_4 is the endogenous substrate with the highest known affinity for this carrier. As indicated by the transport of LTD_4 and N-acetyl-LTE_4 via this ATP-dependent carrier, the glutathione moiety is not a structural determinant of the substrate properties, although it may be a property providing higher affinity for the active site of the carrier. ATP-dependent glutathione S-conjugate transport has been originally described in erythrocyte inside-out membrane vesicles (Kondo et al. 1980) and subsequently observed in other tissues (Kobayashi et al. 1988, 1990; Ishikawa et al. 1989; Akerboom et al. 1991). The carriers expressed in different tissues may be similar in substrate specificity but are distinct as evidenced by the hereditary deficiency of the leukotriene export carrier in liver (Huber et al. 1987) and its simultaneous presence in erythrocytes (Board et al. 1992).

The ATP-dependent leukotriene export carrier in the canalicular membrane is apparently absent or inactive in a mutant strain of rats in which cysteinyl leukotriene excretion into bile is reduced to less than 2% of normal (Huber et al. 1987; Ishikawa et al. 1990). These mutant rats are partially deficient in the hepatobiliary excretion of several other non-bile salt amphiphilic organic anions, such as bilirubin glucuronide and dibromosulfophthalein (Jansen et al. 1985). The defect in this TR⁻ mutant rat strain is considered analogous to the one in Dubin-Johnson syndrome in man and in Corriedale sheep (Jansen et al. 1985; Kitamura et al. 1992). Deficiency of the leukotriene export carrier in the canalicular membrane is compensated by metabolic inactivation and degradation of the leukotrienes in the hepatocyte resulting in an increased renal excretion of leukotriene catabolites (Huber et al. 1987; Keppler et al. 1991).

Inhibition of ATP-dependent transport of LTC_4 in liver is observed in the presence of various glutathione S-conjugates in the micromolar concentration range (Ishikawa et al. 1989). Moreover, structural analogs of LTD_4 and LTE_4, developed as LTD_4/LTE_4 receptor antagonists and devoid of a glutathione moiety and a fatty acid side chain, are not only potent inhibitors of LTC_4 transport in mastocytoma cells (Schaub et al. 1991) but also of the export carrier in the canalicular membrane. Cyclosporin A interferes with the hepatobiliary excretion of cysteinyl leukotrienes (Hagmann et al. 1989). Recent studies demonstrate 50% inhibition of ATP-dependent LTC_4 transport across the rat liver canalicular membrane at a cyclosporin A concentration of 4.5 μM. This inhibition by cyclosporin A is analogous to the inhibition of the ATP-dependent multidrug export carrier (p-glycoprotein) by this immunosuppressant (Foxwell et al. 1989; Speeg et al. 1992).

3.4 Elimination and Transport In Vivo

The pathways of elimination of leukotrienes in the intact organism were originally studied by autoradiographic (Appelgren and Hammarström 1982) and invasive techniques, mostly by use of ^{3}H-labeled leukotrienes (Hagmann et al. 1984; Denzlinger et al. 1985, 1986; Hammarström et al. 1985; Huber et al. 1990; Maltby et al. 1990). Few studies addressed the elimination and in vivo degradation of LTB_4 (Serafin et al. 1984; Hagmann and Korte 1990). More extensive investigations dealt with the elimination of different cysteinyl leukotrienes (Hagmann et al. 1984, 1986; Örning et al. 1985, 1986; Sala et al. 1990; Keppler et al. 1991, 1992). Once released into the blood circulation, the leukotrienes are selectively bound to albumin (Falk et al. 1989) and eliminated predominantly by hepatobiliary excretion. Using N-acetyl-LTE_4 as a representative tracer, half-lives in blood during the initial elimination period were 38 s in the rat and 4 min in man (Keppler et al. 1992). The advantage of using N-acetyl-LTE_4, radioactively labeled in the N-acetylcysteine moiety, is the metabolic stability of the label as opposed to the extensive loss of tritium from leukotrienes labeled in the arachidonate-derived fatty acid moiety during β-oxidation from the ω end (Keppler et al. 1989; Jedlitschky et al. 1991). N-Acetyl-LTE_4 is also an endogenous metabolite of LTC_4 in human urine (Huber et al. 1989; Maltby et al. 1990) and in rodent bile (Hagmann et al. 1986). Moreover, N-acetyl-LTE_4 is eliminated and transported on the same routes and at comparable rates as the other cysteinyl leukotrienes, LTC_4, LTD_4, and LTE_4. For administration of the labeled compound in vivo it is advantageous that the biological activity of N-acetyl-LTE_4 is low when compared to LTD_4 and LTC_4 (Lewis et al. 1981; Samhoun et al. 1989). Within 1 h, 80% of intravenously administered N-acetyl-LTE_4 is excreted in the rat with bile, either intact or after partial oxidative degradation from the ω end of the fatty acid chain (Jedlitschky et al. 1991). At the same time, renal excretion in the rat amounts to about 2%. In man and in the monkey cysteinyl leukotriene excretion into urine represents a much higher proportion than in rodents and amounts to about 50% of the hepatobiliary excretion (Denzlinger et al. 1986; Maltby et al. 1990; Keppler et al. 1992).

Positron emission tomography using carbon-11 labeled, positron-emitting N-[^{11}C]acetyl-LTE_4 enables noninvasive analyses of elimination kinetics, organ distribution, and transport of this cysteinyl leukotriene (Keppler et al. 1991). In the rat, the initial distribution phase was characterized by a rapid disappearance of ^{11}C radioactivity from the blood circulation. This was accompanied by an increase in the leukotriene concentration in liver reaching its maximum 4 min after intravenous injection. In the *cynomolgus* monkey this maximum was reached after 12 min. As a consequence of hepatobiliary excretion, increasing amounts of N-[^{11}C]acetyl-LTE$_4$ and its ω-/β-oxidized metabolites were detected in the intestines. Only negligible amounts of the leukotrienes were monitored in the urinary bladder of the rat within 50 min. Renal excretion was significant, however, in the monkey, which is in accordance with previous invasive tracer studies in this species (Denzlinger et al. 1986). Kinetic analyses indicated a mean transit time of the cysteinyl leukotriene through the liver of 17 min in the rat and of 34 min in the monkey (Keppler et al. 1991). In a mutant rat strain with a hereditary defect of the hepatobiliary transport of cysteinyl leukotrienes across the hepatocyte canalicular membrane (Huber et al. 1987; Ishikawa et al. 1990) elimination of leukotriene radioactivity from the blood circulation was retarded, the mean transit time or storage period in the liver was extended to 54 min, and leukotriene excretion into the intestines was below detectability. This impaired hepatobiliary elimination was compensated by transport of ω-/β-oxidized metabolites from the liver back into blood with subsequent renal excretion. This was monitored by the sharp rise in ^{11}C radioactivity in the urinary bladder of mutant rats. A similar shift from hepatobiliary to renal cysteinyl leukotriene elimination was observed in rats with extrahepatic cholestasis due to surgical ligation of the bile duct. Leukotrienes labeled with a short-lived, positron-emitting radioisotope thus provide quantitative insight into the pathways of their elimination and transport in vivo and into the relative contribution of liver and kidney to these processes under normal and under pathopysiological conditions.

4 Metabolic Deactivation and Inactivation of Leukotrienes

Enzyme-catalyzed chemical modification of the leukotrienes determines their biological activity. Removal of the γ-glutamyl moiety from LTC$_4$ yields the biologically most potent cysteinyl leukotriene, LTD$_4$ (Fig. 2). On the other hand, modification of the cysteinylglycine moiety of LTD$_4$ and ω-oxidation followed by β-oxidation of LTE$_4$, N-acetyl-LTE$_4$, and LTB$_4$ result in deactivation and inactivation of these leukotrienes (Fig. 5). Inactivation of potent mediators is equally important as their biosynthesis since the relative rates of synthesis and inactivation determine the concentration of the biologically active leukotrienes at the receptor.

4.1 Deactivation of LTD_4 in the Mercapturate Pathway

The rank order of molar potencies of the cysteinyl leukotrienes in most assay systems is $LTD_4 > LTC_4 > LTE_4 > N$-acetyl-$LTE_4$ (Lewis et al. 1981; Samhoun et al. 1989). LTD_4 is irreversibly hydrolyzed to LTE_4 and glycine (Fig. 2). The reaction is catalyzed by a dipeptidase (Bernström and Hammarström 1981), which has been purified from microvillus membranes (Kozak and Tate 1982) and characterized as an ectoenzyme (Huber and Keppler 1987). The dipeptidase is associated in the membrane with other enzymes of the mercapturate pathway (Hughey et al. 1978). Removal of the glycine moiety from LTD_4 leads to a considerable loss of biological activity by about two orders of magnitude (Samhoun et al. 1989). Degradation of the glutathione conjugate LTC_4 to LTD_4, LTE_4, and N-acetyl-LTE_4 follows the mercapturate pathway, originally known as a route of detoxification of xenobiotics (Hagmann et al. 1986; Huber and Keppler 1988). In this pathway, the cysteinyl leukotrienes are endogenous substrates in the nanomolar concentration range (Denzlinger et al. 1985). Inhibition of the deactivation of LTD_4 to LTE_4, both on cultured cells and in the rat, is induced by L-penicillamine (Huber and Keppler 1987). This interference also prevents the generation of the mercapturate, N-acetyl-LTE_4, and of ω-oxidized polar metabolites of LTE_4 and N-acetyl-LTE_4.

N-Acetyl-LTE_4 is formed by intracellular N-acetylation of LTE_4 with acetyl-coenzyme A (CoA). The enzyme catalyzing this reaction is present in liver, kidney, spleen, skin, and lung of the rat (Bernström and Hammarström 1986). Endogenous N-acetyl-LTE_4 was originally identified in rat bile (Hagmann et al. 1985, 1986) and feces (Örning et al. 1986) as the predominant LTC_4 catabolite. In human urine, but not in the bile, this mercapturate is present as a minor metabolite amounting to about 10% of LTE_4 (Huber et al. 1989, 1990; Sala et al. 1990; Maltby et al. 1990). N-acetyl-LTE_4 retains at least 30% of activity relative to LTE_4 (Lewis et al. 1981) and may be equipotent as LTE_4 in some assays (Samhoun et al. 1989). Therefore, catabolism of LTC_4 in the mercapturate pathway is associated with biological activation to LTD_4, followed by partial deactivation to LTE_4 and N-acetyl-LTE_4. Complete inactivation of the cysteinyl leukotrienes is only achieved by oxidation at the ω end of the fatty acid moiety (Samhoun et al. 1989).

4.2 Oxidative Inactivation of Leukotrienes

ω-Oxidation of LTB_4 to ω-hydroxy-LTB_4, ω-aldehyde-LTB_4, and ω-carboxy-LTB_4, which is associated with a reduction of biological activity, has

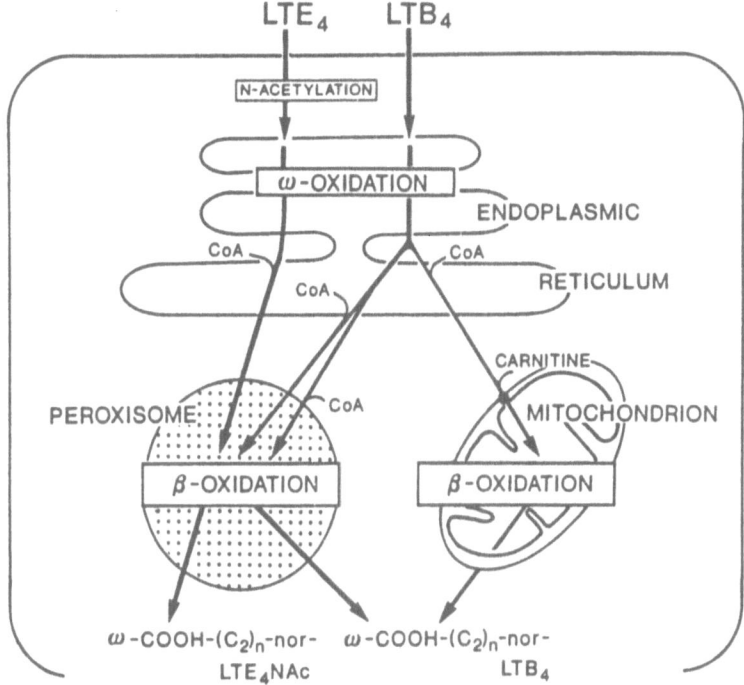

Fig. 5. Compartmentation of inactivation and degradation of LTE_4 and LTB_4 in the hepatocyte. Products of peroxisomol β-oxidation may be dinor, tetranor, and hexanor metabolites of the LTE_4, N-acetyl-LTE_4, or LTB_4 [ω-COOH-$(C_2)_n$-nor-leukotriene]. (Reproduced with permission from Jedlitschky et al. 1991)

been observed in leukocytes (Hansson et al. 1981; Powell 1984; Soberman et al. 1988; Lewis et al. 1990) as well as in hepatocytes (Harper et al. 1986; Baumert et al. 1989; Shirley and Murphy 1990; Sumimoto et al. 1990; Jedlitschky et al. 1990; Shirley et al. 1992). By identification of ω-carboxy-dinor-LTB_4 and ω-carboxy-tetranor-LTB_3 in hepatocyte suspensions, these cells were shown to β-oxidize ω-carboxy-LTB_4 from the ω end (Harper et al. 1986; Jedlitschky et al. 1991). Ethanol at moderate concentrations interferes with the further catabolism of ω-hydroxy-LTB_4 (Baumert et al. 1989). As a result, not only LTB_4 and ω-hydroxy-LTB_4 (Baumert et al. 1989) but also 3-hydroxy-LTB_4 increases in hepatocytes (Shirley et al. 1992). The latter are potent calcium-mobilizing and chemotactic metabolites (Shirley et al. 1992).

The liver converts LTE_4 and N-acetyl-LTE_4 to the respective ω-hydroxy and ω-carboxy metabolites (Örning 1987; Ball and Keppler 1987; Stene and Murphy 1988). Further degradation by β-oxidation from the ω end yields ω-carboxy-dinor, -tetranor, and -hexanor derivates of LTE_4

and N-acetyl-LTE$_4$ (Stene and Murphy 1988; Sala et al. 1990; Huber et al. 1990). All these ω-carboxy derivates of LTE$_4$ and N-acetyl-LTE$_4$ are biologically inactive (Samhoun et al. 1989).

Additional pathways for the catabolism of cysteinyl leukotrienes have been described on the basis of in vitro experiments. These include the degradation of LTC$_4$, LTD$_4$, and LTE$_4$ to 6-*trans*-LTB$_4$ diastereoisomers and the subclass-specific S-diastereoisomeric sulfoxides by myeloperoxidase from activated human polymorphonuclear leukocytes and monocytes (Lee et al. 1983). Additionally, cysteinyl leukotrienes may be inactivated by hydroxyl radicals yielding 6-*trans*-isomers of LTB$_4$ (Henderson et al. 1982). These pathways for inactivation of cysteinyl leukotrienes have been outlined repeatedly (Lewis et al. 1990), however, there is no evidence for their significance in the intact organism where the metabolites recovered from injected LTC$_4$, LTD$_4$, LTE$_4$, or N-acetyl-LTE$_4$ in bile and urine account for most of the administered leukotrienes and exclude a detectable contribution from myeloperoxidase-catalyzed degradation in vivo (Huber et al. 1987; Jedlitschky et al. 1991).

4.3 Peroxisomal Degradation by β-Oxidation from the ω End

The peroxisomal β-oxidation pathway for very long-chain fatty acids involves acyl-CoA oxidase (EC 1.3.99.3), the bifunctional or rather trifunctional protein displaying enoyl-CoA hydratase (EC 4.2.1.17), 3-hydroxyacyl-CoA dehydrogenase (EC 1.1.1.35), and Δ^3, Δ^2-enoyl-CoA isomerase (EC 5.3.3.8) activity, and the peroxisomal 3-ketoacyl-CoA thiolase (EC 2.3.1.16), as well as auxiliary enzymes such as 2,4-dienoyl-CoA reductase (EC 1.3.1.34; Osmundsen et al. 1991). The increased degradation of leukotrienes in the β-oxidation pathway after treatment of rats with clofibrate, an inducer of peroxisome proliferation, led to the suggestion that β-oxidation of leukotrienes may be localized in peroxisomes (Keppler et al. 1989). Both the long-chain structure of the leukotrienes and the structures of their degradation products by β-oxidation in rat hepatocytes (Shirley and Murphy 1990) and in human urine (Sala et al. 1990; Huber et al. 1990) are in line with peroxisomal leukotriene breakdown. It is of interest that the leukotrienes, in contrast to the prostaglandins, are not degraded from the carbon-1-carboxyl group but from the ω end by β-oxidation of the ω-carboxy metabolites derived from LTB$_4$, LTE$_4$, and N-acetyl-LTE$_4$. Direct evidence for an exclusive degradation of cysteinyl leukotrienes in peroxisomes has been obtained by use of isolated liver peroxisomes and direct photoaffinity labeling of the peroxisomal enzymes of β-oxidation with ω-carboxy-N-[^{3}H]acetyl-LTE$_4$ (Jedlitschky et al. 1991). In addition, isolated peroxisomes catalzye the β-oxidation from the ω end of ω-carboxy-LTB$_4$ yielding the dinor and the tetranor catabolites (Jedlitschky et al. 1991). In vitro experiments indicate that the degradation of LTB$_4$ can also proceed in liver mitochondria, as indicated in Fig. 5. It is unlikely, however, that the mitochondrial β-oxidation of ω-carboxy-LTB$_4$ plays a major role in the intact organism since LTB$_4$ degradation is severly impaired in patients with Zellweger syndrome, a disorder of peroxisomal biogenesis (Mayatepek et al. 1992). In this inherited disease the defect of peroxisomal leukotriene degradation results in increased levels of the biologically active, proinflammatory mediators LTE$_4$ and LTB$_4$. In addition, the concentrations in urine of ω-carboxy-LTE$_4$ and ω-carboxy-LTB$_4$, which are the immediate substrates for peroxisomal β-oxidation, are manifold increased (Mayatepek et al. 1992). These findings in humans with peroxisome deficiency underline the essential role of peroxisomes in the catabolism of leukotrienes.

5 Analysis of Cysteinyl Leukotrienes and LTB$_4$

Quantitative determinations of the leukotrienes can be accomplished by radioimmunoassays, high-performance liquid chromatography (HPLC), mass spectrometry after gas chromatography, bioassays, or combinations of these techniques. Separation of leukotriene metabolites by HPLC (Borgeat et al. 1990) often serves as initial step prior to detection with high sensitivity. For unequivocal identification gas chromatography/mass spectrometry is the method of choice (Murphy 1984; Mathews 1990; Murphy and Sala 1990).

5.1 Methods for Determination in Biological Fluids

Difficulties in leukotriene analysis include (a) the short half-life of these mediators in vivo and in most biological fluids, (b) their presence in low nanomolar or picomolar concentration, (c) their susceptibility to oxidative degradation during sample preparation, and (d) the artificial generation of leukotrienes from cells during sampling particularly the leukotriene release from blood leukocytes during attempts to measure blood plasma leukotrienes (Denzlinger et al. 1986).

Most measurements of LTB$_4$ in biological fluids have employed sensitive radioimmunoassays after verification of the identity of substance by HPLC or mass spectrometry (Tateson et al. 1988; Lehr et al. 1991; Mayatepek et al. 1992).

The endogenous cysteinyl leukotrienes have been analyzed in fluids into which these substances are excreted and present at sufficient concentrations, particularly in bile (Hagmann et al. 1984, 1985; Denzlinger et al. 1985, 1986; Keppler 1988) and urine (Denzlinger et al. 1986; Keppler et al. 1988; Huber et al. 1989; Tagari et al. 1989; Taylor et al. 1989; Nicoll-Griffith et al. 1990; Denzlinger et al. 1990; Fauler et al. 1991). These determinations have been based on tracer studies which have defined species-characteristic index metabolites for systemic cysteinyl leukotriene production (Keppler et al. 1988). In humans, the measurement of urinary LTE_4 reflects about 5% of the systemic LTC_4 production (Maltby et al. 1990); in the rat, N-acetyl-LTE_4 represents the index metabolite of choice to be analyzed in bile and corresponding to about 13% of systemic LTC_4 generation (Huber and Keppler 1987) and in the guinea pig biliary LTD_4 amounts to 20%–50% of LTC_4 administered into the systemic blood circulation (A. Keppler et al. 1987; Guhlmann et al. 1989). In each case, HPLC separation of the respective index metabolite in urine or bile should precede the quantitative analysis by immunoassay or mass spectrometry, and the results should be corrected for the recovery of internal standards. The percentages of cysteinyl leukotrienes eliminated into bile and urine are influenced by the relative transport capacities of these organs as well as by the enzyme activities degrading the cysteinyl leukotrienes in the vascular bed, hepatocytes, and kidney. Nevertheless, these determinations provide useful information on the role of leukotrienes under pathophysiological conditions and on the action of inhibitors of their synthesis, whereas analyses in blood are less meaningful because of the short half-life of these mediators in the systemic circulation and the risk of their artificial ex vivo synthesis and release from blood cells (Denzlinger et al. 1986; Heavy et al. 1987; Keppler 1988).

5.2 Generation of Cysteinyl Leukotrienes In Vivo
Under Pathophysiological Conditions

Pathophysiological conditions associated with enhanced systemic generation of cysteinyl leukotrienes have been described in experimental animals and in humans. In most instances, local release of the mediators leads to elimination with the blood circulation followed by biliary and renal excretion of detectable quantities. Under a few experimental conditions, such as in the anaphylactic shock in the guinea pig (A. Keppler et al. 1987; Guhlmann et al. 1989), a causal relationship has been established between the quantitiy of leukotriene release and the clinical symptoms.

In humans, biliary LTE_4 is increased in acute pancreatitis (Keppler 1988). Enhanced urinary LTE_4 excretion is associated with fulminant hepatitis, liver cirrhosis, and hepatorenal syndrome (Huber et al. 1989; Moore et al. 1990), antigen challenge in asthma patiens (Taylor et al. 1989; Tagari et al. 1989, 1990; Christie et al. 1991), adult respiratory distress syndrome and burns (Fauler et al. 1991; Westcott et al. 1991), systemic lupus erythematosus (Hackshaw et al. 1992), and treatment with certain cytokines, such as granulocyte-macrophage colony-stimulating factor (Denzlinger et al. 1990), and tumor necrosis factor-α. In addition to urinary LTE_4, both LTE_4 and LTB_4 have been determined in significant quantities in sputum from patients with cystic fibrosis and asthma (Piper et al. 1991). In the monkey, intoxication with staphylococcal enterotoxin B (Denzlinger et al. 1986) and endotoxin from *Salmonella abortus equi* elicited increased biliary and urinary LTE_4 excretion.

Systemic anaphylaxis leads to an immediate release of relatively large amounts of cysteinyl leukotrienes detected as LTD_4 in guinea pig bile (A. Keppler et al. 1987; Guhlmann et al. 1989), or as N-acetyl-LTE_4 in rat bile (Foster et al. 1988), or as LTE_4 in sheep lymph during cyclooxygenase blockade (Robinson et al. 1986). Biliary cysteinyl leukotrienes also increase after immunological challenge of the isolated rat or guinea pig liver (Hagmann et al. 1991).

In the rat, where 85%–90% of the systemic LTC_4 production is reflected by the biliary excretion of metabolites, various pathophysiological conditions have been studied by analysis of N-acetyl-LTE_4 in bile. These disease states include endotoxin shock (Hagmann et al. 1984, 1985, 1986; D. Keppler et al. 1987), different types of tissue trauma such as surgical trauma, bone fracture, burn injury (Denzlinger et al. 1985), shock induced by platelet-activating factor (Huber and Keppler 1987) and by tumor necrosis factor-α (Huber et al. 1988), and fulminant experimental hepatitis (Hagmann et al. 1987).

6 Leukotriene-Mediated Disease Processes and Their Prevention

The biological actions of LTC_4, LTD_4 LTE_4, and LTB_4, as well as actions of some of the metabolites, such as N-acetyl-LTE_4 and ω-hydroxy-LTB_4, have been well defined (Dahlén et al. 1981; Lewis and Austen 1984; Piper 1984; Feuerstein 1985; Drazen and Austen 1987; Samuelsson et al. 1987; Guhlmann et al. 1989; Rola-Pleszczynski 1989; Ford-Hutchinson 1990; Lehr et al. 1991; Shaw and Krell 1991; Shirley et al. 1992). Moreover, analysis of leukotriene concentrations in biological fluids and tissues have established that the concentrations and amounts of these mediators under some conditions are sufficient to elicit pathophysiological responses in humans and experimental animals (A. Keppler et al. 1987; Keppler et al. 1988; Guhlmann et al. 1989; Taylor et al. 1989). Of particular importance are the recent results from clinical studies with receptor antagonists and inhibitors of leukotriene biosynthesis. The development of both types of compounds has reached a high degree of selectivity in their actions and sufficient bioavailability (Gillard et al. 1989; Aharony and Krell 1991; Ford-Hutchinson 1991b; Jones et al. 1991; Lewis et al. 1991; Piper and Krell 1991; Herron et al. 1992). Thus, the criteria to define and the means to treat leukotriene-mediated disease processes are available. Anaphylactic shock in the sensitized guinea pig may serve as an example where selective inhibition of leukotriene biosynthesis in vivo (by MK-886) prevents the generation of otherwise lethal amounts of endogenous LTC_4, and where the above mentioned criteria have been fulfilled (Guhlmann et al. 1989).

Asthma is the human disease in which the most convincing evidence has been presented to implicate the cysteinyl leukotrienes as key mediators (Piper and Krell 1991; Lewis et al. 1991). This conclusion is based on the bronchoconstrictor activity of inhaled LTC_4, LTD_4, and LTE_4 (Drazen and Austen 1987), on the generation of cysteinyl leukotrienes during the asthmatic attack (Taylor et al. 1989), and on the results from clinical studies with third-generation LTD_4/LTE_4 receptor antagonists (Lewis et al. 1991). In other inflammatory diseases, local or systemic leukotriene production has been measured and suggests a role in pathogenesis; however, the importance remains to be proven by successful clinical intervention or prevention by use of leukotriene biosynthesis inhibitors, LTD_4/LTE_4 receptor antagonists, and/or LTB_4 receptor antagonists. The leukotrienes may act within a network of mediators involving cytokines and other arachidonate metabolites. Diseases in which inhibition of leukotriene synthesis or action may prove to be beneficial include, in addition to asthma and anaphylaxis, psoriasis, adult respiratory distress syndrome, neonatal pulmonary hypertension, allergic rhinitis, gout, rheumatoid arthritis, inflammatory bowel disease, acute and fulminant hepatitis, hepatorenal syndrome, glomerulonephritis, and possibly sepsis. The use of selective leukotriene synthesis inhibitors and receptor antagonists may result not only in therapeutic progress but also in a deeper and more detailed understanding of the role of leukotrienes under normal and pathophysiological conditions.

References

Aharony D, Krell RD (1991) Pharmacology of peptide leukotriene receptor antanonists. Ann NY Acad Sci 629:125–132

Akerboom TPM, Narayanaswami V, Kunst M, Sies H (1991) ATP-dependent S-(2,4-dinitrophenyl)glutathione transport in canalicular plasma membrane vesicles from rat liver. J Biol Chem 266:13147–13152

Appelgren LE, Hammarström S (1982) Distribution and metabolism of [^{3}H]-labeled leukotriene C_3 in the mouse. J Biol Chem 257:531–535

Augstein J, Farmer JB, Lee TB, Sheard P, Tattersall ML (1973) Selective inhibition of slow reacting substance of anaphylaxis. Nature New Biol 245:214–216

Ball HA, Keppler D (1987) ω-oxidation products of leukotriene E_4 in bile and urine of the monkey. Biochem Biophys Res Commun 148:664–670

Baumert T, Huber M, Mayer D, Keppler D (1989) Ethanol-induced inhibition of leukotriene degradation by ω-oxidation. Eur J Biochem 182:223–229

Bell RL, Young PR, Albert D, Lanni C, Summers JB, Brooks DW, Rubin P, Carter GW (1992) The discovery and development of zileuton – an orally active 5-lipoxygenase inhibitor. Int J Immunopharmacol 14:505–510

Bernström K, Kammarström S (1981) Metabolism of leukotriene D by porcine kidney. J Biol Chem 256:9579–9582

Bernström K, Kammarström S (1986) Metabolism of leukotriene E_4 by rat tissues: formation of N-acetyl leukotriene E_4. Arch Biochem Biophys 244:486–491

Board P, Nishida T, Gatmaitan Z, Che M, Arias IM (1992) Erythrocyte membrane transport of glutathione conjugates and oxidized glutathione in the Dubin-Johnson syndrome and in rats with hereditary hyperbilirubinemia. Hepatology 15:722–725

Borgeat P, Picard S, Vallerand P, Bourgoin S, Odeimat A, Sirois P, Poubelle PE (1990) Automated on-line extraction and profiling of lipoxygenase products of arachidonic acid by high-performance liquid chromatography. Meth Enzymol 187:98–116

Christie PE, Tagari P, Ford-Hutchinson AW, Charleson S, Chee P, Arm JP, Lee TH (1991) Urinary leukotriene-E_4 concentrations increase after aspirin challenge in aspirin-sensitive asthmatic subjects. Am Rev Resp Dis 143:1025–1029

Dahinden CA, Wirthmueller U (1990) Release and metabolism of leukotriene A_4 in neutrophil-mast cell interactions. Methods Enzymol 187:567–577

Dahinden CA, Clancy RM, Gross M, Chiller JM, Hugli TE (1985) Leukotriene C_4 production by murine mast cells: evidence of a role for extracellular leukotriene A_4. Proc Natl Acad Sci USA 82:6632–6636

Dahlén SE, Björck T, Hedqvist P, Arfors KE, Hammarström S, Lindgren JÄ, Samuelsson B (1981) Leukotrienes promote plasma leakage and leukocyte adhesion in postcapillary venules: in vivo effects with relevance to the acute inflammatory response. Proc Natl Acad Sci USA 80:3887–3891

Denzlinger C, Rapp S, Hagmann W, Keppler D (1985) Leukotrienes as mediators in tissue trauma. Science 230:330–332

Denzlinger C, Guhlmann A, Scheuber PJ, Wilker D, Hammer DK, Keppler D (1986) Metabolism and analysis of cysteinyl leukotrienes in the monkey. J Biol Chem 261:15601–15606

Denzlinger C, Kapp A, Grimberg M, Gerhartz HH, Wilmanns W (1990) Enhanced endogenous leukotriene biosynthesis in patients treated with granulocyte-macrophage colony-stimulating factor. Blood 76:1765–1770

Dixon RAF, Jones RE, Diehl RE, Bennett CD, Kargman S, Rouzer CA (1988) Cloning of the cDNA for human 5-lipoxygenase. Proc Natl Acad Sci USA 85:416–420

Dixon RAF, Diehl RE, Opas E, Rands E, Vickers PJ, Evans JF, Gillard JW, Miller DK (1990) Requirement of a 5-lipoxygenase-activating protein for leukotriene synthesis. Nature 343:282–284

Drazen JM, Austen KF(1987) Leukotrienes and airway responses. Am Rev Respir Dis 136:985–998

Evans JF, Léveillé C, Mancini JA, Prasit P, Thérien M, Zamboni R, Gauthier JY, Fortin R, Charleson P, MacIntyre DE, Luell S, Bach TJ, Meurer R, Guay J, Vickers PJ, Rouzer CA, Gillard JW, Miller DK (1991) 5-Lipoxygenase-activating protein is the target of a quinoline class of leukotriene synthesis inhibitors. Molec Pharmacol 40:22–27

Falk E, Müller M, Huber M, Keppler D, Kurz G (1989) Direct photoaffinity labeling of leukotriene binding sites. Eur J Biochem 186:741–747

Fauler J, Tsikas D, Holch M, Seekamp A, Nerlich ML, Sturm J, Frölich JC (1991) Enhanced urinary excretion of leukotriene E_4 by patients with multiple trauma with or without adult respiratory distress syndrome. Clin Sci 80:497–504

Feinmark SJ (1990) Leukotriene C_4 biosynthesis during polymorphonuclear leukocyte-vascular cell interactions. Methods Enzymol 187:559–567

Ferber E, Resch K (1973) Phospholipid metabolism of stimulated lymphocytes: activation of acyl-CoA: lysolecithin acyltransferases in microsomal membranes. Biochim Biophys Acta 296:335–349

Feuerstein G (1985) Autonomic pharmacology of leukotrienes. J Auton Pharmac 5: 149–168

Fitzsimmons BJ, Rockach J (1989) Enzyme inhibitors and leukotriene receptor antagonists. In: Roçkach J (ed) Leukotrienes and lipoxygenases. Chemical, biological and clinical aspects. Elsevier, Amsterdam, pp 427–502

Ford-Hutchinson AW (1990) Leukotriene B_4 in inflammation. Crit Rev Immunol 10:1–12

Ford-Hutchinson AW (1991a) FLAP: a novel drug target for inhibiting the synthesis of leukotrienes. TIPS 121:68–70

Ford-Hutchinson AW (1991b) Leukotrienes and chemotaxis – 5-lipoxygenase activation and control. In: Sies H, Flohé L, Zimmer G (eds) Molecular aspects of inflammation. Springer, Berlin Heidelberg New York, pp 33–39

Foster A, Letts G, Charleson S, Fitzsimmons B, Blacklock B, Rokach J (1988) The in vivo production of peptide leukotrienes after pulmonary anaphylaxis in the rat. J Immunol 141:3544–3550

Foxwell BM, Mackie A, Ling V, Ryffel B (1989) Identification of the multidrug resistance-related P-glycoprotein as a cyclosporine binding protein. Mol Pharmacol 36:543–546

Gillard J, Ford-Hutchinson AW, Chan C, Charleson S, Denis D, Foster A, Fortin R, Leger S, McFarlane CS, Morton H, Piechuta H, Riendeau D, Rouzer CA, Rokach J, Young R, MacIntyre DE, Peterson L, Bach T, Eiermann G, Hopple S, Humes J, Hupe L, Luell S, Metzger J, Meurer R, Miller DK, Opas E, Pacholok S (1989) L-663,536 (MK-886) (3-[1-(4-chlorbenzyl)-3-t-butyl-thio-5-isopropylindol-2-yl]-2,2-dimethylpropanoic acid), a novel orally active leukotriene biosynthesis inhibitor. Can J Physiol Pharmacol 67:456–464

Guhlmann A, Keppler A, Kästner S, Krieter H, Brückner UB, Messmer K, Keppler D (1989) Prevention of endogenous leukotriene production during anaphylaxis in the guinea pig by an inhibitor of leukotriene biosynthesis (MK-886) but not by dexamethasone. J Exp Med 170:1905–1918

Habenicht AJR, Goerig M, Rothe DER, Specht E, Ziegler R, Glomset JA, Graf T (1989) Early reversible induction of leukotriene synthesis in chicken myelomonocytic cells transformed by a temperature sensitive mutant of avian leukemia virus E26. Proc Natl Acad Sci USA 86:921–924

Hackshaw KV, Voelkel NF, Thomas RB, Westcott JY (1992) Urine leukotriene E_4 levels are elevated in patients with active systemic lupus erythematosus. J Rheumatol 19: 252–258

Haeggström JR, Wetterholm A, Vallee BL, Samuelsson B (1990) Leukotriene A_4 hydrolase: an epoxide hydrolase with peptidase activity. Biochem Biophys Res Commun 173: 431–437

Hagmann W, Korte M (1990) Hepatic uptake and metabolic disposition of leukotriene B_4 in rats. Biochem J 267:467–470

Hagmann W, Denzlinger C, Keppler D (1984) Role of peptide leukotrienes and their hepatobiliary elimination in endotoxin action. Circ Shock 14:223–235

Hagmann W, Denzlinger C, Keppler D (1985) Production of peptide leukotrienes in endotoxin shock, FEBS Lett 180:309–313

Hagmann W, Denzlinger C, Rapp S, Weckbecker G, Keppler D (1986) Identification of the major endogenous leukotriene metabolite in the bile of rats as N-acetyl leukotriene E_4. Prostaglandins 31:239–251

Hagmann W, Steffan A-M, Kirn A, Keppler D (1987) Leukotrienes as mediators in frog virus 3-induced hepatitis in rats. Hepatology 7:732–736

Hagmann W, Parthé S, Kaiser I (1989) Uptake, production and metabolism of cysteinyl leukotrienes in the isolated perfused rat liver. Inhibition of leukotriene uptake by cyclosporine. Biochem J 261:611–616

Hagmann W, Kaiser I, Jakschik BA (1991) The sensitized liver represents a rich source of endogenous leukotrienes. Hepatology 13:482–488

Hammarström S (1983) Leukotrienes. Ann Rev Biochem 52:355–377

Hammarström S, Örning L, Bernström K (1985) Metabolism of leukotrienes. Mol Cell Biochem 69:7–16

Hansson G, Lindgren J.-Ä, Dahlén SE, Hedqvist P, Samuelsson B (1981) Identification and biological activity of novel omega-oxidized metabolites of leukotriene B_4 from human leukocytes. FEBS Lett 130:107–112

Harper TW, Garrity MJ, Murphy RC (1986) Metabolism of leukotriene B_4 in isolated rat hepatocytes. Identification of a novel 18-carboxy-19,20-dinor leukotriene B_4 metabolite. J Biol Chem 261:5414–5418

Heavey DJ, Soberman RJ, Lewis RA, Spur B, Austen KF (1987) Critical considerations in the development of an assay for sulfidopeptide leukotrienes in plasma. Prostaglandins 33:693–708

Henderson WR, Jörg A, Klebanoff SJ (1982) Eosinophil peroxidase-mediated inactivation of leukotrienes B_4, C_4 and D_4. J Immunol 128:2609–2613

Herron DK, Goodson T, Bollinger NG, Swansonbean D, Wright IG, Staten GS, Thompson AR, Froelich LL, Jackson WT (1992) Leukotriene-B_4 receptor antagonists – the LY255283-series of hydroxyacetophenones. J Med Chem 35:1818–1828

Huber M, Keppler D (1987) Inhibition of leukotriene D_4 catabolism by D-penicillamine. Eur J Biochem 167:73–79

Huber M, Keppler D (1988) Leukotrienes and the mercapturate pathway. In: Sies H, Ketterer B (eds) Glutathione conjugation. Mechanisms and biological significance. Academic, London, pp 449–470

Huber M, Keppler D (1990) Eicosanoids an the liver. In: Popper H, Schaffner F (eds) Progress in liver diseases, vol 9. Saunders, Philadelphia, pp 117–141

Huber M, Guhlmann A, Jansen PLM, Keppler D (1987) Hereditary defect of hepatobiliary cysteinyl leukotriene elimination in mutant rats with defective hepatic anion excretion. Hepatology 7:224–228

Huber M, Beutler B, Keppler D (1988) Tumor necrosis factor α stimulates leukotriene production in vivo. Eur J Immunol 18:2085–2088

Huber M, Kästner S, Schölmerich J, Gerok W, Keppler D (1989) Cysteinyl leukotriene analysis in human urine: enhanced excretion in patients with liver cirrhosis and hepatorenal syndrome. Eur J Clin Invest 19:53–60

Huber M, Müller J, Leier I, Jedlitschky G, Ball HA, Moore KP, Taylor GW, Williams R, Keppler D (1990) Metabolism of cysteinyl leukotrienes in monkey and man. Eur J Biochem 194:309–315

Hughey RP, Rankin BB, Elce JS, Curthoys NP (1978) Specificity of a particulate rat renal peptidase and its localization along with other enzymes of mercapturic acid synthesis. Arch Biochem Biophys 186:211–217

Irvine RF (1982) How is the level of free arachidonic acid controlled in mammalian cells? Biochem J 204:3–16

Ishikawa T, Kobayashi K, Sogame Y, Hayashi K (1989) Evidence for leukotriene C_4 transport mediated by an ATP-dependent glutathione S-conjugate carrier in rat heart and liver plasma membranes. FEBS Lett 259:95–98

Ishikawa T, Müller M, Klünemann C, Schaub T, Keppler T (1990) ATP-dependent primary active transport of cysteinyl leukotrienes across liver canalicular membrane. Role of the ATP-dependent transport system for glutathione S-conjugates. J Biol Chem 265: 19279–19286

Jansen PLM, Peters WH, Lamers WH (1985) Hereditary chronic conjugated hyperbilirubinemia in mutant rats caused by defective hepatic anion transport. Hepatology 5:573–579

Jedlitschky G, Leier I, Huber M, Mayer D, Keppler D (1990) Inhibition of leukotriene ω-oxidation by ω-trifluoro analogs of leukotrienes. Arch Biochem Biophys 282:333–339

Jedlitschky G, Huber M, Völkl A, Müller M, Leier I, Müller J, Lehmann W-D, Fahimi HD, Keppler D (1991) Peroxisomal degradation of leukotrienes by β-oxidation from the ω-end. J Biol Chem 266:24763–24772

Jones DA, Fitzpatrick FA (1990) Leukotriene B_4 biosynthesis by erythrocyte-neutrophil interactions. Methods Enzymol 187:553–559

Jones TR, Zamboni R, Belley M, Champion E, Charette L, Ford-Hutchinson AW, Gauthier J-Y, Leger S, Lord A, Masson P, McFarlane CS, Metters KM, Pickett C, Piechuta H, Young RN (1991) Pharmacology of the leukotriene antagonist verlukast: the (R)-enantiomer of MK-571. Can J Physiol Pharmacol 69:1847–1854

Keppler A, Örning L, Bernström K, Hammarström S (1987) Endogenous leukotriene D_4 formation during anaphylactic shock in the guinea pig. Proc Natl Acad Sci USA 84: 5903–5907

Keppler D (1988) Stoffwechsel und Analyse von Leukotrienen in Vivo. Klin Wochenschr. 66:997–1004

Keppler D, Hagmann W, Rapp S, Denzlinger C, Koch HK (1985) The relation of leukotrienes to liver injury. Hepatology 5:883–891

Keppler D, Hagmann W, Rapp S (1987) Role of leukotrienes in endotoxin action in vivo. Rev Infect Diseases 9:580–584

Keppler D, Huber M, Hagmann W, Ball HA, Guhlmann A, Kästner S (1988) Metabolism and analysis of endogenous cysteinyl leukotrienes. Ann NY Acad Sciences 524:68–74

Keppler D, Huber M, Baumert T, Guhlmann A (1989) Metabolic inactivation of leukotrienes. Adv Enzyme Regul 28:307–319

Keppler D, Guhlmann A, Oberdorfer F, Krauss K, Müller J, Ostertag H, Huber M (1991) Generation and metabolism of cysteinyl leukotrienes in vivo. Ann NY Acad Sciences 629:100–104

Keppler D, Müller M, Klünemann C, Guhlmann A, Krauss K, Müller J, Berger U, Leier I, Mayatepek E (1992) Transport and in vivo elimination of cysteinyl leukotrienes. Adv Enzyme Regul 32:107–116

Kitamura T, Alroy J, Gatmaitan Z, Inoue M, Mikami T, Jansen P, Arias IM (1992) Defective biliary excretion of epinephrine metabolites in mutant (TR⁻) rats: relation to the pathogenesis of black liver in the Dubin-Johnson syndrome and Corriedale sheep with an analogous excretory defect. Hepatology 15:1154–1159

Kobayashi K, Sogame Y, Hayashi K, Nicotera P, Orrenius S (1988) ATP stimulates the uptake of S-dinitrophenylglutathione by rat liver plasma membrane vesicles. FEBS Lett 240:55–58

Kobayashi K, Sogame Y, Hara H, Hayashi K (1990) Mechanism of glutathione S-conjugate transport in canalicular and basolateral rat liver plasma membranes. J Biol Chem 265:7737–7741

Kondo T, Dale GL, Beutler E (1980) Glutathione transport by inside-out vesicles from human erythrocytes. Proc Natl Acad Sci USA 77:6359–6362

Kozak EM, Tate SS (1982) Glutathione-degrading enzymes of microvillus membranes. J Biol Chem 257:6322–6327

Kurz G, Müller M. Schramm U, Gerok W (1989) Identification and function of bile salt binding polypeptides of hepatocyte membrane. In: Petzinger E, Kinne RK-H, Sies H (eds) Hepatic transport of organic substances. Springer, Berlin Heidelberg New York, pp 267–278

Lam BK, Owen WF Jr, Austen KF, Soberman RJ (1989) The identification of a distinct export step following the biosynthesis of leukotriene C_4 by human eosinophils. J Biol Chem 264:12885–12889

Lam BK, Gagnon L, Austen KF, Soberman RJ (1990) The mechanism of leukotriene B_4 export from human polymorphonuclear leukocytes. J Biol Chem 265:13438–13441

Lee CW, Lewis RA, Tauber AI, Mehrotra M, Corey EJ, Austen KF (1983) The myeloperoxidase-dependent metabolism of leukotrienes C_4 D_4 and E_4 to 6-*trans*-leukotriene B_4 diastereoisomers and the subclass-specific *S*-diastereoisomeric sulfoxides. J Biol Chem 258:15004–15010

Lehr HA, Guhlmann A, Nolte D, Keppler D, Messmer K (1991) Leukotrienes as mediators in ischemia-reperfusion injury in a microcirculation model in the hamster. J Clin Invest 87:2036–2041

Leier I, Müller M. Jedlitschky G, Keppler D (1992) Leukotriene uptake by hepatocytes and hepatoma cells. Eur J Biochem 209: in press

Lewis MA, Krell RD, Jones TR (1991) Third-generation peptidoleukotriene receptor antagonists. In: Crooke ST, Wong A (eds) Lipoxygenases and their products. Academic, San Diego, pp 207–234

Lewis RA, Austen KF (1984) The biologically active leukotrienes. Biosynthesis, metabolism, receptors, functions, and pharmacology. J Clin Invest 73:889–897

Lewis RA, Drazen JM, Austen KF, Toda M, Brion F, Marfat A, Corey EJ (1981) Contractile activities of structural anlogs of leukotrienes C and D: role of the polar substituents. Proc Natl Acad Sci USA 78:4579–4583

Lewis RA, Austen KF, Soberman RJ (1990) Leukotrienes and other products of the 5-lipoxygenase pathway. Biochemistry and relation to pathobiology in human diseases. New Engl J Med 323:645–655

Lin L-L, Lin AY, Knopf JL (1992) Cytosolic phospholipase A_2 is coupled to hormonally regulated release of arachidonic acid. Proc Natl Acad Sci USA 89:6147–6151

Maltby NH, Taylor GW, Ritter JM, Moore K, Fuller RW, Dollery CT (1990) Leukotriene C_4 elimination and metabolism in man. J Allergy Clin Immunol 85:3–9

Mathews R (1990) Quantitative gas chromatography-mass spectrometry analysis of leukotriene B_4. Meth Enzymol 187:76–81

Matsumoto T, Funk CD, Radmak O, Höög JO, Jörnvall H, Samuelsson B (1988) Molecular cloning and amino acid sequence of human 5-lipoxygenase. Proc Natl Acad Sci USA 85: 26–30

Mayatepek E, Lehmann W-D, Fauler J, Tsikas D, Frölich JC, Schutgens RBH, Wanders RJA, Keppler D (1992) Impaired degradation of leukotrienes in patients with peroxisome deficiency disorders. J Clin Invest, in press

McGee JE, Fitzpatrick F (1985) Enzymatic hydration of leukotriene A_4: purification and characterization of a novel epoxide hydrolase from human erythrocytes. J Biol Chem 260:12832–12837

McGee JE, Fitzpatrick FA (1986) Erythrocyte-neutrophil interactions: formation of leukotriene B_4 by transcellular biosynthesis. Proc Natl Acad Sci USA 83:1349–1353

Miller DK, Gillard JW, Vickers PJ, Sadowski S, Léveillé C, Mancini JA, Charleson P, Dixon RAF, Ford-Hutchinson AW, Fortin R, Gauthier JY, Rodkey J, Rosen R, Rouzer CA, Sigal IS, Strader CD, Evans JF (1990) Identification and isolation of a membrane protein necessary for leukotriene production. Nature 343:278–281

Minami M, Ohishi N, Mutoh H, Izumi T, Bito H, Wada H, Seyama Y, Toh H, Shimizu T (1990) Leukotriene A_4 hydrolase is a zinc-containing aminopeptidase. Biochem Biophys Res Commun 173:620–626

Moore KP, Taylor GW, Maltby NH, Siegers D, Fuller RW, Dollery CT, Williams R (1990) Increased production of cysteinyl leukotrienes in hepatorenal syndrome. J Hepatol 11:263–271

Müller M, Ishikawa T, Berger U, Lucka L, Klünemann C, Schreyer A, Kannicht C, Reutter W, Kurz G, Keppler D (1991a) ATP-dependent transport of taurocholate across the hepatocyte canalicular membrane mediated by a 110 kDa glycoprotein binding ATP and bile salt. J Biol Chem 266:18920–18926

Müller M, Falk E, Sandbrink R, Berger U, Leier I, Jedlitschky G, Huber M. Kurz G, Keppler D (1991b) Photoaffinity labeling of leukotriene binding sites in hepatocytes and hepatoma cells. Advan Prostaglandin, Thromboxane, Leukotriene Res 21A:395–398

Murphy RC (1984) Mass spectrometric quantitation and analysis of leukotrienes and other 5-lipoxygenase metabolites. Prostaglandins 28:597–601

Murphy RC, Sala A (1990) Quantitation of sulfidopeptide leukotrienes in biological fluids by gas chromatography-mass spectrometry. Meth Enzymol 187:90–98

Murphy RC, Hammarström S, Samuelsson B (1979) Leukotriene C: a slow-reacting substance from murine mastocytoma cells. Proc Natl Acad Sci USA 76:4275–4279

Nicoll-Griffith D, Zamboni R (1990) BIO-fully automated sample treatment high-performance liquid chromatography and radioimmunoassay for leukotriene E_4 in human urine from asthmatics. J Chromatography 526:341–354

Odlander B, Jakobsson PJ, Rosén A, Claesson H-E (1988) Human B and T lymphocytes convert leukotriene A_4 into leukotriene B_4. Biochem Biophys Res Commun 153: 203–208

Örning L (1987) ω-Oxidation of cysteine-containing leukotrienes by rat liver microsomes. Isolation and characterization of ω-hydroxy and ω-carboxy metabolites of leukotriene E_4 and N-acetyl-leukotriene E_4. Eur J Biochem 170:77–85

Örning L, Kaijser RL, Hammarström S (1985) In vivo metabolism of leukotriene C_4 in man: Urinary excretion of leukotriene E_4. Biochem Biophys Res Commun 130:214–220

Örning L, Norin E, Gustafsson B, Hammarström S (1986) In vivo metabolism of leukotriene C_4 in germ-free and conventional rats. Fecal excretion of N-acetyl-leukotriene E_4. J Biol Chem 261:766–771

Örning L, Krivi G, Bild G, Gierse J, Aykent S, Fitzpatrick FA (1991a) Inhibition of leukotriene A_4 hydrolase/aminopeptidase by captopril. J Biol Chem 266:16507–16511

Örning L, Krivi G, Fitzpatrick FA (1991b) Leukotriene A_4 hydrolase. Inhibition by bestatin and intrinsic aminopeptidase activity establish its functional resemblance to metallohydrolase enzymes. J Biol Chem 266:1375–1378

Ormstad K, Uehara N, Orrenius S, Örning L, Hammarström S (1982) Uptake and metabolism of leukotriene C_4 by isolated rat organs and cells. Biochem Biophys Res Commun 104:1434–1440

Osmundsen H, Bremer J, Pedersen JI (1991) Metabolic aspects of peroxisomal β-oxidation. Biochim Biophys Acta 1085:141–158

Peppelenbosch MP, Tertoolen LGJ, Den Hertog J, De Laat SW (1992) Epidermal growth factor activates calcium channels by phospholipase A_2/5-lipoxygenase-mediated leukotriene C_4 production. Cell 69:295–303

Piper PJ (1984) Formation and actions of leukotrienes. Physiol Rev 64:744–761

Piper PJ, Krell RD (eds) (1991) Advances in the understanding and treatment of asthma. Ann NY Acad Sciences 629:89–167

Piper PJ, Conroy DM, Costello JF, Evans JM, Green CP, Price JF, Sampson AP, Spencer DA (1991) Leukotrienes and inflammatory lung disease. Ann NY Acad Sciences 629: 112–119

Powell WS (1984) Properties of leukotriene B_4 hydrolase from polymorphonuclear leukocytes. J Biol Chem 259:3082–3089

Reid GK, Kargman S, Vickers PJ, Mancini JA, Léveillé C, Ethier D, Miller DK, Gillard JW, Dixon RAW, Evans JF (1990) Correlation between expression of 5-lipoxygenase-activating protein, 5-lipoxygenase, and cellular leukotriene synthesis. J Biol Chem 265: 19818–19823

Robinson DR, Skoskiewicz M, Bloch JK, Castorena G, Hayes E, Lowenstein E, Melvin C, Michelassi F, Zapol WM (1986) Cyclooxygenase blockade elevates leukotriene E_4 production during acute anaphylaxis in sheep. J Exp Med 163:1509–1517

Rokach J (ed) (1989) Leukotrienes and lipoxygenases. Chemical, biological and clinical aspects. Elsevier, Amsterdam

Rola-Pleszczynski M (1989) Leukotrienes and the immune system. J Lipid Mediators 1: 149–159

Rouzer CA, Kargman S (1988) Translocation of 5-lipoxygenase to the membrane in human leukocytes challenged with ionophore A23187. J Biol Chem 263:10980–10988

Rouzer CA, Rands E, Kargman S, Jones RE, Register RB, Dixon RAF (1988) Characterization of cloned human leukocyte 5-lipoxygenase expressed in mammalian cells. J Biol Chem 263:10135–10140

Rouzer CA, Ford-Hutchinson AW, Morton HE, Gillard JW (1990) MK886, a potent and specific leukotriene biosynthesis inhibitor blocks and reverses the membrane association of 5-lipoxygenase in ionophore-challenged leukocytes. J Biol Chem 265:1436–1442

Samhoun MN, Conroy RM, Piper PJ (1989) Pharmacological profile of leukotriene E_4, N-acetyl E_4 and four of their novel ω- and β-oxidative metabolites in airways of guinea-pig and man in vitro. Br J Pharmacol 98:1406–1412

Samuelsson B (1983) Leukotrienes: mediators of immediate hypersensitivity reactions and inflammation. Science 220:568–575

Samuelsson B, Funk CD (1989) Enzymes involved in the biosynthesis of leukotriene B_4. J Biol Chem 264:19469–19472

Samuelsson B, Borgeat P, Hammarström S, Murphy RC (1979) Introduction of a nomenclature: leukotrienes. Prostaglandins 17:785–787

Samuelsson B, Dahlén S-E, Lindgren JÅ, Rouzer CA, Serhan CN (1987) Leukotrienes and lipoxins: structures, biosynthesis, and biological effects. Science 237:1171–1176

Saussy DL Jr, Sarau HM, Mong S, Crooke ST (1989) Mechanisms of leukotriene E_4 partial agonist activity at leukotriene D_4 receptors in differentiated U-937 cells. J Biol Chem 264:19845–19855

Schaub T, Ishikawa T, Keppler D (1991) ATP-dependent leukotriene export from mastocytoma cells. FEBS Lett 279:83–86

Serafin WE, Oates JA, Hubbard WC (1984) Metabolism of leukotriene B_4 in the monkey. Identification of the principal nonvolatile metabolite in the urine. Prostaglandins 27: 899–911

Shaw A, Krell RD (1991) Peptide leukotrienes: current status of research. J Medicinal Chem 34:1235–1242

Shirley MA, Murphy RC (1990) Metabolism of leukotriene B_4 in isolated rat hepatocytes. Involvement of 2,4-dienoyl-coenzyme A reductase in leukotriene B_4 metabolism. J Biol Chem 265:16288–16295

Shirley MA, Reidhead CT, Murphy RC (1992) Chemotactic LTB_4 metabolites produced by hepatocytes in the presence of ethanol. Biochem Biophys Res Commun 185:604–610

Snyder DW, Fleisch JK (1989) Leukotriene receptor antagonists as potential therapeutic agents. Annu Rev Pharmacol Toxicol 29:123–143

Soberman RJ, Sutyak JP, Okita RT, Wendelborn DF, Roberts II LJ, Austen KF (1988) The identification and formation of 20-aldehyde leukotriene B_4. J Biol Chem 263:7996–8002

Söderström M, Hammarström S, Mannervik B (1988) Leukotriene C synthase in murine mastocytoma cells: a novel enzyme distinct from cystolic and microsomal glutathione transferase. Biochem J 250:713–718

Sola A, Voelkel N, Maclouf J, Murphy RC (1990) Leukotriene E_4 elimination and metabolism in normal human subjects. J Biol Chem 265:21771–21778

Speeg KV, Maldonado AL, Liaci J, Muirhead D (1992) Effect of cyclosporine on colchicine secretion by a liver canalicular transporter studied in vivo. Hepatology 15:899–903

Stene DO, Murphy RC (1988) Metabolism of leukotriene E_4 in isolated rat hepatocytes. Identification of β-oxidation products of sulfidopeptide leukotrienes. J Biol Chem 263: 2773–2778

Sumimoto H, Kusunose E, Gotoh Y, Kusunose M, Minakami S (1990) Leukotriene B_4 ω-hydroxylase in rat liver microsomes: identification as a cytochrome P-450 that catalyzes prostaglandin A_1 ω-hydroxylation, and participation of cytochrome b_5. J Biochem 108:215–221

Tagari P, Ethier D, Carry M, Korley V, Charleson S, Girard Y, Zamboni R (1989) Measurement of urinary leukotrienes by reversed-phase liquid chromatography and radioimmunoassay. Clin Chem 35:388–391

Tagari P, Rasmussen JB, Delorme D, Girard Y, Eriksson L-O, Charleson S, Ford-Hutchinson AW (1990) Comparison of urinary leukotriene E_4 and 16-carboxytetranordihydro leukotriene E_4 excretion in allergic asthmatics after inhaled antigen. Eicosanoids 3:75–80

Tateson JE, Randall RW, Reynolds CH, Jackson WP, Bhattacherjee P, Salmon JA, Garland LG (1988) Selective inhibition of arachidonate 5-lipoxygenase by novel acetohydroxamic acids: biochemical assessment in vitro and ex vivo. Br J Pharmacol 94:528–539

Taylor GW, Taylor I, Black P, Maltby NH, Fuller RW, Dollery CT (1989) Urinary leukotriene E_4 after antigen challenge and in acute asthma and allergic rhinitis. Lancet i:584–589

Uehara N, Ormstad K, Örning L, Hammarström S (1983) Characteristics of the uptake of cysteine-containing leukotrienes by isolated hepatocytes. Biochim Biophys Acta 732:69–74

Verhagen J, Bruynzeel PLB, Koedam JA, Wassink GA, de Boer M, Terpstra GK, Kreukniet J, Veldink GA, Vliegenthart JFG (1984) Specific leukotriene formation by purified human eosinophils and neutrophils. FEBS Lett 168:23–28

Weckbecker G, Keppler D (1986) Leukotriene C_4 metabolism by hepatoma cells deficient in the uptake of cysteinyl leukotrienes. Eur J Biochem 154:559–562

Westcott JY, Thomas RB, Voelkel NF (1991) Elevated urinary leukotriene E_4 in patients with ARDS and severe burns. Prostaglandins Leukotrienes and Essential Fatty Acids 43:151–158

Wettstein M, Gerok W, Häussinger D (1989) Metabolism of cysteinyl leukotrienes in nonrecirculating rat liver perfusion. Eur J Biochem 181:115–124

Yoshimoto T, Soberman RJ, Spur B, Austen KF (1988) Properties of highly purified leukotriene C_4 synthase of guinea pig lung. J Clin Invest 81:866–871

Rev. Physiol. Biochem. Pharmacol., Vol. 121

The Hypothalamic Hormone Oxytocin:
From Gene Expression to Signal Transduction

EVITA MOHR, WOLFGANG MEYERHOF, and DIETMAR RICHTER

Contents

1 Introduction

Since the discovery of neuropeptides the biology and biochemistry of these substances have been extensively investigated in many laboratories. It is now firmly established that peptide hormones play a key role both in the central nervous system and in peripheral endocrine systems in the regulation of physiology and behavior, consistent with their wide distribution throughout the animal kingdom. Although our knowledge about the function of the peptidergic neuron is accumulating, basic questions concerning gene evolution, the regulation of gene expression, protein modification, protein targeting, and signal transduction remain to be fully answered.

Institut für Zellbiochemie und klinische Neurobiologie, UKE, Universität Hamburg, D-2000 Hamburg 20, FRG

In this review, these questions will be addressed to the biology of oxytocin, which is one of the best characterized peptides. As early as the beginning of this century it was noted that pituitary extracts stimulated uterus contraction and milk ejection in the experimental animal (Dale 1906; Ott and Scott 1910), now known to be the classical endocrine functions of oxytocin. A few years later the first evidence was presented that neurons of the hypothalamus possess all the characteristics of a secretory cell, which enabled and stimulated research in the biosynthesis of oxytocin and many other hormones in the peptidergic neuron of the central nervous system (Scharrer and Scharrer 1940; Bargmann 1966).

2 Biosynthesis

The classical site of oxytocin biosynthesis is represented by the hypothalamopituitary tract of the central nervous system, namely in magnocellular neurons of the supraoptic and paraventricular nuclei. Most of the axons of these cells project via the median eminence and the neural stalk to the posterior pituitary, where the peptide is released into the blood stream destined for endocrine functions in the periphery. Some cells of the paraventricular nucleus, however, project their axons into other brain regions (Swanson and Sawchenko 1983) where oxytocin is thought to act as either a neurotransmitter or a modulator. Oxytocin-containing cells have also been identified in a number of peripheral tissues, indicative of hormone synthesis outside the brain; these include the adrenal gland, corpus luteum, and placenta (for review see Ivell 1986 and references cited therein).

For historical reasons the biosynthesis of oxytocin has always been associated with that of the closely related antidiuretic hormone vasopressin. Immunocytochemical experiments, however, showed that the two peptides are present within different subsets of magnocellular neurons in the paraventricular and supraoptic nuclei (Dierickx and Vandesande 1979).

Like many other hormones, oxytocin is synthesized as a composite polyprotein precursor, the preprohormone (Gainer et al. 1977a). A signal peptide which directs protein synthesis to the lumen of the endoplasmic reticulum (Walter and Blobel 1981) is followed by the hormone moiety and a third protein component called neurophysin. To release the biologically active peptide, the precursor has to undergo a series of posttranslational processing events involving the sequential action of several proteolytic enzymes. After cotranslational removal of the signal peptide, the prohormone is endoproteolytically cleaved at a dibasic signal, thereby liberating oxytocin which is extended by GlyXX (X = basic amino acid). Characterization of dibasic processing endopeptidases is still in its infancy but it seems clear that these enzymes belong to the subtilisin family of serine proteases which were initially believed to be restricted to prokaryotes (for review see Barr 1991). Subtilisin-like proteases have been detected in lower eukaryotes such as yeast (Fuller et al. 1989). Subsequently, cDNAs encoding several dibasic processing enzymes have been isolated from a variety of mammalian species by virtue of their extensive sequence similarities to their bacterial and yeast counterparts (Wise et al. 1990; Barr et al. 1991). Although definitive proof is still lacking, there is growing evidence that these proteases are also responsible for cleavage at dibasic residues in vivo (Barr 1991).

Following endoproteolytic cleavage, the carboxy-terminal amino acids are removed by carboxypeptidase E (Clamigirand et al. 1987). This enzyme has been purified from a number of tissues and its activity is widely distributed throughout the central nervous system and in peripheral organs (Fricker and Snyder 1983; Fricker et al. 1986).

Frequently, neuropeptides, including oxytocin, contain a carboxyterminal α-amide group that is important for biological activity and which contributes to its stability in plasma (for review see Bradbury and Smyth 1991). Amides are formed by an enzyme-catalyzed reaction, with glycine serving as the nitrogen donor (Bradbury et al. 1982). This, in fact, requires the activity of two different enzymes and, thus, is performed in two steps. In the first reaction a hydroxyglycine intermediate is generated by peptidyl glycine hydroxylase, which requires ascorbate and copper. In the second step the peptide amide and glyoxylic acid are formed by the activity of a peptidylhydroxyglycine N-C lyase. Both enzymes together are commonly referred to as peptide-amidating enzyme (PAM).

Comparisons of the amino acid sequences of oxytocin-associated neurophysins obtained from several species by Edmann degradation (Chauvet et al. 1983) and molecular cloning techniques (Rehbein et al. 1986) reveal that the polypeptide deduced from the respective cDNA sequence contains a supernumerary basic amino acid at its C terminus, probably representing a rudimentary processing signal, which is cleaved off by a carboxypeptidase B-like enzyme (Nörenberg and Richter 1988).

The processing and modification of the oxytocin precursor takes place during its transport in neurosecretory vesicles from the perikarya to the nerve terminals in the posterior pituitary (Gainer et al. 1977b). Following depolarization, the peptide content is released in a Ca^{2+}-dependent manner.

3 Gene Structure and Evolution

With the advent of modern molecular biological techniques, the gene
structures and genomic organizations of the oxytocin and the closely related vasopressin genes have been determined in a variety of species
(Schmale et al. 1983; Ivell and Richter 1984; Ruppert et al. 1984; Sausville
et al. 1985; Hara et al. 1990). Both genes are composed of three exons, encoding the signal peptide, the hormone moiety and the N-terminal part of
neurophysin (exon A), the central part of neurophysin (exon B), and the C-
terminal part of neurophysin as well as, in the vasopressin precursor, a
glycopeptide (exon C). In all mammalian species examined so far the two
genes are closely linked on the same chromosome (Fig. 1) and are oriented
in opposite transcriptional directions (Hara et al. 1990; Mohr et al. 1988;
Sausville et al. 1985).

Recent data obtained by cDNA cloning in nonmammalian vertebrates
have shed some light on the evolution of the oxytocin/vasopressin gene
family. The organization in the teleost fish *Catostomus commersoni* of the
genes encoding the homologous hormones named isotocin and vasotocin,
respectively, is very similar to that found in the mammal. However, in the
fish, the neurophysin moieties of both hormone precursors are extended by
about 30 amino acid residues. Although these segments show sequence
similarities to the mammalian glycopeptide referred to as the copeptin of
the vasopressin precursor, they do not contain any glycosylation or appropriate processing signals (Heierhorst et al. 1989; Morley et al. 1990). It is
likely that, during the course of evolution, this extended part of neuro-

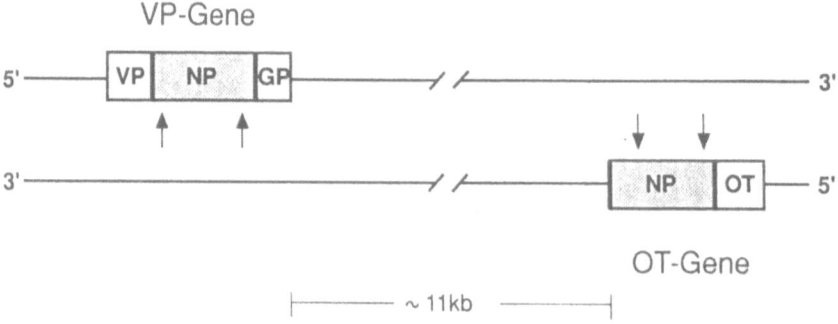

Fig. 1. Structural organization of the rat vasopressin and oxytocin genes. The exons are indicated by *boxes* and show: *VP*, vasopressin; *NP*, neurophysin; *OT*, oxytocin; *GP*, glycopeptide. The *arrows* indicate the relative positions of the intervening sequences. The genes
are separated by about 11 kilobases (*kb*) of intergenic sequences and are oriented in opposite transcriptional directions

physin developed into a separate protein component of distinct function (the copeptin in the case of the vasopressin precursor) or in the case of the oxytocin precursor was deleted because of lack of function.

4 Gene Expression and Regulation

In a variety of physiological conditions such as late gestation, parturition, and lactation the peripheral demand for oxytocin is high. Accordingly, precursor synthesis in and hormone release from magnocellular neurons is accelerated. A similar response is also seen during dehydration, although a role for oxytocin in the maintenance of salt and water balance remains obscure (Forsling and Brimble 1985; van Tol et al. 1987).

When these studies were extended to the level of gene expression, it became evident that a high precursor synthesis is tightly coupled to elevated levels of oxytocin mRNA (Forsling 1986; van Tol et al. 1987, 1988; Zingg and Lefebvre 1988). However, definite proof is still lacking as to whether or not mRNA content rises because of a higher rate of gene transcription or is the consequence of a longer transcript half-life. Because of the close relationship of the vasopressinergic and oxytocinergic systems in the hypothalamus, the expression of both genes is commonly looked at. Whenever the levels of oxytocin gene transcripts are elevated, higher levels of vasopressin-encoding mRNA are observed and vice versa. Since these peptides serve quite different functions in the organism, these findings remain an enigma. Yet, quantitative northern blot analyses have shown that oxytocin and vasopressin mRNAs rise two- to three-fold during late gestation, and sustained elevated levels of both transcripts are also observed during lactation (Zingg and Lefebvre 1988; van Tol et al. 1988).

The influence of ovarian steroid hormones on the hypothalamic oxytocinergic system is well documented (Akaishi and Sakuma 1985; Negoro et al. 1973; Yamaguchi et al. 1979). The observed parallel rise in estrogen and oxytocin mRNA levels during lactation and birth (Yoshinaga et al. 1969) suggests a direct influence of the steroid hormone on oxytocin gene expression. Several lines of evidence supported this view. For example, a concomitant variation in the oxytocin mRNA and estrogen levels has been observed during the rat estrous cycle (van Tol et al. 1988) and ovariectomy is accompanied by a fall in oxytocin transcript levels (Miller et al. 1989). The most compelling line of evidence was provided by the detection of two estrogen response elements (ERE) in the promoter region of the rat oxytocin gene, which turned out to be functional enhancers in a heterologous system (Fig. 2; Mohr and Schmitz 1991). However, earlier in situ autoradiography studies failed to detect estrogen-concentrating

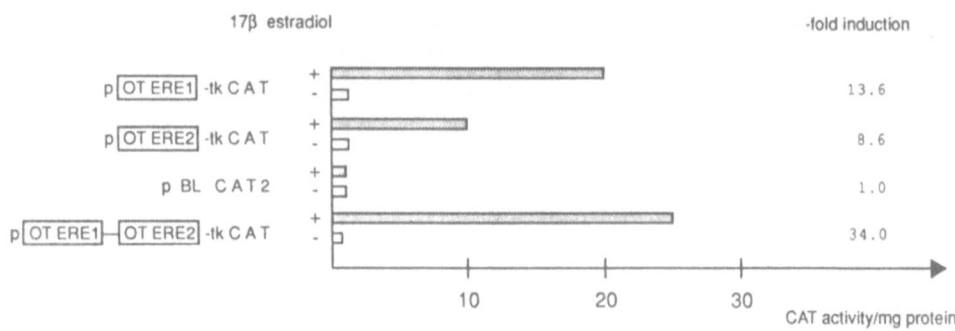

Fig. 2. Determination of CAT activities in cell lysates of MCF 7 cells transiently transfected with either various OT ERE tk CAT constructs or the parent vector pBL CAT2 with (+) or without (−) the addition of 17β-estradiol to the culture medium. The *lower part* shows a schematic diagram of the relative positions and nucleotide sequences of OT ERE1 and OT ERE2 in the rat oxytocin gene promoter region. *TATA*, modified TATA box; *arrowhead*, transcription initiation site (+1); *ERE*, estrogen response element; *OT*, oxytocin; *tk*, Herpes simplex thymidine kinase gene promotor; *CAT*, chloramphenicol acetyltransferase. For experimental details see Mohr and Schmitz (1991)

cells in magnocellular neurons projecting to the neurohypophysis, indicating the absence there of estrogen receptors (Rhodes et al. 1981). It is likely that estrogen receptor-mediated transcriptional activation does not occur in cells of the hypothalamo-neurohypophyseal tract which are responsible for oxytocin release into the circulation. However, estrogen target cells projecting into brain regions other than the neural lobe have been detected in the posterior part of the paraventricular nucleus (Rhodes et al. 1981). Oxytocin synthesized in these cells is thought to function either as a neurotransmitter or a modulator (Swanson and Sawchenko 1983). Accordingly, in these neurons oxytocin gene expression might be triggered by estrogens at the transcriptional level.

Further attemps to define regulatory elements in the oxytocin gene promotor that direct cell- and tissue-specific gene expression have been unsuccessful up to now because of the lack of suitable cell culture systems. Recent investigations of rat oxytocin gene expression in transgenic mice demonstrated that cell-specific transcription of the transgene only occurs in magnocellular neurons if the oxytocin gene is linked to the vasopressin gene in a minilocus (Young III et al. 1990), thus mimicking the in vivo situation. These data suggest the localization of oxytocin-specific enhancers either within or in the vicinity of the vasopressin gene. Additional constructs will be needed to define the putative *cis*-acting elements more precisely. In contrast, the specific expression of the vasopressin gene has not been observed in these animals.

5 Axonal Transport of Oxytocin mRNA

The complexity and plasticity of the peptidergic neuron is reflected by the recent finding that oxytocin mRNA, as well as vasopressin mRNA, is axonally transported from the perikarya of magnocellular neurons to the nerve terminals in the neural lobe (Mohr et al. 1991). Although it is generally believed that axons contain neither mRNA nor organelles involved in protein biosynthesis (Lasek and Brady 1981; Gordon-Weeks 1988), experimental evidence is now growing to suggest that axons of both vertebrates and invertebrates contain mRNA (Giuditta et al. 1977; Koenig 1979; Dirks et al. 1989; Giuditta et al. 1990; van Minnen and Schallig 1990; Giuditta et al. 1991). However, clues to the function and physiological significance of these transcripts are currently lacking. One attractive hypothesis is to assume that these mRNAs are locally translated in the axon. Due to the apparent absence of rough endoplasmic reticulum and Golgi apparatus in the nerve endings, translation would have to take place on free ribosomes or polysomes. Moreover, the precursor polyproteins are probably not subjected to posttranslational processing and modification and are likely secreted by a mechanism that differs completely from the classical mode of neuropeptide release, i.e., via neurosecretory vesicles. It seems unlikely that axonal mRNA simply reflects an overflow from the cell bodies, since the process of selective mRNA transport into the dendrites of neurons is well known (Davis et al. 1987; Tucker et al. 1989).

6 Physiological Effects and Binding Sites for Oxytocin

A detailed understanding of the role of oxytocin as an important hormone in the periphery and an authentic neuromodulator in the central nervous system requires a correlation of its physiological effects with cellular actions and with the existence of specific high-affinity binding sites in the appropriate tissues, the characterization or purification of a receptor, and the identification of a coupled signal transduction system. We will now summarize the current evidence for this association.

Circulating oxytocin that is released from the neurohypophysis or from peripheral tissues elicits a variety of physiological effects in the periphery. The best characterized examples of this are its function in female reproduction in a number of mammalian species including humans, i.e., in uterine contractions during the onset and maintenance of parturition, and the process of milk let-down during lactation which is mediated by the myoepithelial cells of mammillary alveolar glands (Johnson and Everitt 1984). In male reproduction oxytocin, apparently synthesized within the testis (Nicholson et al. 1984), is involved in contraction of the smooth muscle of the vas deferens, the epididymis, and the seminiferous tubules during ejaculation (reviewed in Wathes 1984). Other roles of oxytocin in biological reproduction include metabolic effects on reproductive tissues (Lederis et al. 1985; Okabe et al. 1985), an immediate release of arachidonic acid and prostaglandin $F_{2\alpha}$ from decidua cells (Wilson et al. 1988), and gonadotropin release from dispersed anterior pituitary cells (Evans and Catt 1989). Moreover, it has been associated with the regulation of renal function (Verbalis et al. 1991). The various physiological responses to oxytocin are thought to be mediated by specific oxytocin receptors, although there is some indication that the neurohypophyseal hormones do not absolutely discriminate their cognate receptors (Jard et al. 1987 and references cited therein; Teitelbaum 1991). The assumption that oxytocin effects are caused by its cognate receptor is underscored by the presence of high-affinity binding sites for oxytocin that are clearly pharmacologically distinct from those for vasopressin in most, if not all, tissues known to respond to the peptide, such as the reproductive tissues from a number of species (Jard et al. 1987), the kidney (Stoeckel et al. 1987; Schmidt et al. 1991), kidney-derived LLC-PK1 cells (Stassen et al. 1988), adenohypophysis (Antoni 1986; Chadio and Antoni 1989), and the thymus (Elands et al. 1990).

A growing body of evidence indicates that oxytocin may function as an authentic neuromodulator in the central nervous system, being involved in drug addiction, ethanol tolerance, learning and memory, sexual and maternal behavior (reviewed in Kovacs 1986), and inhibition of food intake (Olson et al. 1991). Support for the neuromodulatory role of oxytocin comes also from studies of its cellular actions. For instance, extracellular recordings indicated that it was able to excite most, but not all, neurons of the dorsal motor nucleus of the vagus nerve and of the bed nucleus of the stria terminalis in slices from rat brainstem (Charpak et al. 1984; Ingram et al. 1990) and to increase the firing rate of nonpyramidal neurons in slices from the CA1 field of the rat and guinea pig hippocampus (Raggenbass et al. 1989). The role for oxytocin in the central nervous system is also highlighted by the observation of high-affinity binding sites in synaptic membrane preparations from various brain regions (Audigier and Barberis

1985). Furthermore, quantitative light microscopic autoradiography and ligand-binding assays have demonstrated specific binding sites for oxytocin in the hypothalamus, in some forebrain regions including the hippocampus, in the olfactory system, the limbic system, the brainstem, certain parts of the striatum and cortex, the neurohypophysis and the ependyma of the lateral ventricle, and/or the choroid plexus near the lateral septum (Tribollet et al. 1988; Di Scala-Guenot et al. 1990). In regions where the binding of oxytocin and arginine vasopressin overlap, the use of a highly specific oxytocin receptor ligand, which show only negligible affinity for the V_1 and V_2 vasopressin receptors clearly established the presence of specific oxytocin binding sites (Elands et al. 1988).

7 Signal Transduction Mediated by Oxytocin

Uterotonic agents such as oxytocin, platelet-activating factor, and norepinephrine are believed to stimulate uterine smooth muscle contraction by increasing the intracellular calcium concentration (Kursheed and Sanborn 1989). Furthermore, it has been shown that inositol 1,4,5 trisphosphate (IP_3) in this tissue mediates the release of calcium from microsomes (Carsten and Miller 1985), suggesting that IP_3 is the intracellular second messenger. This assumption is supported by the detection of myometrial IP_3 binding sites with characteristics of true IP_3 receptors that display a high apparent dissociation constant, fast association and dissociation rates, a correct binding profile of inositol phosphate isomers, and an inability to metabolize IP_3 (Rivera et al. 1990).

A number of studies have indicated that oxytocin binding to myometrial tissues from a number of species (Schrey et al. 1986; Marc et al. 1986), cultured rat inner medullary collecting tubule cells (Teitelbaum 1991), and dispersed bovine mammary cells (Zhao and Gorewit 1987) results in phospholipase C (PLC)-mediated phosphatidylinositol bisphosphate hydrolysis and activate the IP_3 – calcium signalling pathway. The presumed activation of phospholipases in the oxytocin signaling pathway is supported by the observation of prostaglandin synthesis and release from the endometrium and from decidua cells (Roberts et al. 1976; Wilson et al. 1988), pointing to the generation of arachidonic acid either by phospholipase A_2 or by the sequential action of PLC and diacylglycerol lipase (Schrey et al. 1986), an enzyme present at high levels in human decidua cells (Okazaki et al. 1981). The involvement of a guanine nucleotide-binding protein (G protein) distinct from G_s and G_i in oxytocin receptor–effector coupling is suggested from studies showing that fluoroaluminates mimic the oxytocin-mediated production of inositol phosphates and myometrial contraction in a pertussis- and cholera-toxin insensitive fashion (Marc et al. 1988). Finally, the oxytocin-dependent stimulation of protein kinase C, which is typical of the inositol phosphate – calcium signaling pathway has been shown to occur in rat adipocytes (Egan et al. 1990). The rise in the oxytocin-induced intracellular calcium concentration may also be due, at least in part, to an inhibition of a rat myometrial plasma membrane (Ca^{2+} + Mg^{2+}) ATPase (Soloff and Sweet 1982) and of a human myometrial sarcolemmal calcium ATPase (Popescu et al. 1985) and/or by an influx of extracellular Ca^{2+} into LLC-PK1 cells (Stassen et al. 1988). Figure 3 schematically outlines the oxytocin-induced signal transduction pathway. In contrast to the situation in the myometrium in the mouse anococcygeus muscle the oxytocin-dependent elevation of cytosolic free Ca^{2+} and contractions were lost in a calcium-free solution, indicating a dominant role for a calcium influx mechanism (Gibson and Shaw 1989).

The coupling of oxytocin receptors to the IP_3–calcium signaling pathway is further supported by functional expression studies in frog oocytes. In the past, this system has frequently been used to express a variety of G protein-coupled receptors either from cellular mRNA or from RNA

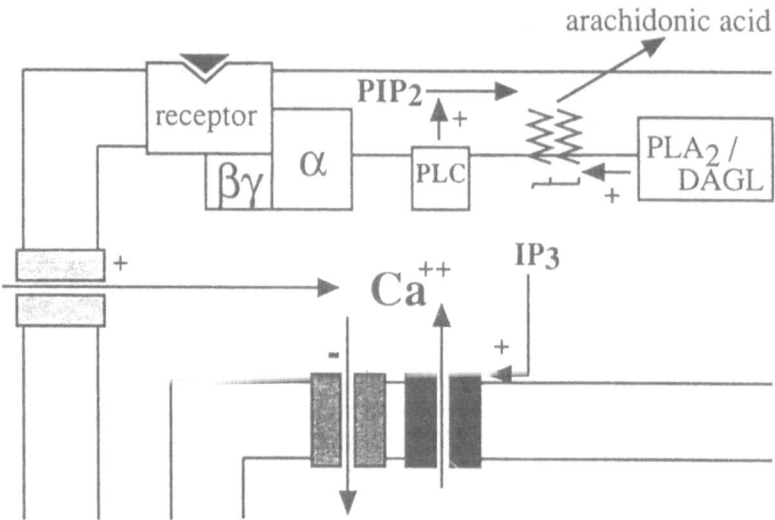

Fig. 3. Oxytocin(OT)-mediated signal transduction pathways. Binding of OT (*black triangle*) to its receptor results in G protein (α and $\beta\gamma$ subunits are indicated) and phosphoinositol-specific phospholipase C activation, causing IP_3-mediated Ca^{2+} release from internal stores via an IP_3 receptor (*black*). Additionally, calcium influx via plasma membrane calcium channels (shown in *light gray*) is enhanced, and sequestration of cytosolic calcium is inhibited through a blockade of the sarcolemmal (Mg^{2+}-Ca^{2+})ATPase (*gray*). Diacylglycerol is thought to activate protein kinase C and is itself hydrolyzed by phospholipase A_2 or diacylglycerol lipase to generate arachidonic acid, the substrate for prostaglandin synthesis: PIP_2, phosphatidylinositol bisphosphate; *DAGL*, diacylglycerol lipase; PLA_2, phospholipase A_2. For details see text

transcribed in vitro from cloned cDNA (for a review see Richter et al. 1991). It has been noted that only those receptors can be detected in voltage-clamp recordings which are linked to the IP_3–calcium response (Meyerhof et al. 1988). Expression of the oxytocin receptor from bovine endometrial mRNA has been achieved in this system (Morley et al. 1988). mRNA from LLC-PK1 cells has also been used in functional expression studies in oocytes by demonstrating oxytocin-mediated $^{45}Ca^{2+}$ efflux (Cantau et al. 1990).

To date, changes in cyclic AMP levels in the uterotonic, galactobolic, lipolytic, and renal actions of oxytocin have been excluded (Teitelbaum 1991; Cantau et al. 1990; Jard et al. 1987) and there is, as far as we are aware, no indication of a coupling of oxytocin responses to adenylyl cyclase.

8 Oxytocin Receptors

Although the oxytocin receptor has yet not been purified, nor its cDNA cloned, the above-cited evidence strongly suggests that it belongs to the well-known class of type II receptors that have seven putative membrane spanning domains and which are coupled to G proteins. It is also currently unclear as to whether distinct oxytocin receptor subtypes mediate the various peripheral and central effects. This question is particularly intriguing because oxytocin receptors can be distinguished from at least three vasopressin receptor subtypes (V_{1a}, V_{1b}, V_2; Jard et al. 1986) and because even highly specific oxytocin analogs for one species show a less pronounced selectivity in another (Tence et al. 1990).

To date, the existence of oxytocin receptor subtypes has not yet been established because the lack of precise pharmacological and functional criteria does not permit such a distinction to be made (Jard et al. 1987). However, there is some circumstantial evidence for oxytocin receptor heterogeneity in the rat myometrium (el Alj et al. 1990) and suggestions that the uterine receptors may differ from those mediating milk ejection (Sawyer et al. 1981). Particularly interesting is the question of whether the central and peripheral effects are mediated by one or two classes of receptors. Pharmacological evidence indicating that a hippocampal oxytocin receptor is of the uterine type (Mühlethaler et al. 1983) must await biochemical data that assigns to the central receptors the second messenger pathway that is activated by the peripheral receptors.

Biochemical studies have revealed that ^3H–oxytocin binding sites can be solubilized from rat mammary plasma membrane preparations, and gel filtration has indicated that the detergent-solubilized binding site has multiple molecular mass forms ranging from 40 kDa to more than 200 kDa (Soloff and Fernstrom 1987). Radiation inactivation analysis pointed to an apparent relative molecular mass of 57.5 +/– 3.8 kDa (Soloff et al. 1988). In photoaffinity labeling experiments and subsequent sodium dodecyl sulfate gel electrophoresis using membrane preparations from rat mammary gland, the size of the oxytocin receptor was determined to be 65 +/– 3 kDa (Müller et al. 1989), which agrees reasonably well with the radiation inactivation study. However, using a photoaffinity labeling approach, a 78 +/– 5-kDa protein was specifically labeled in membrane preprations from guinea pig myometrium (Fahrenholz et al. 1988). The size of an oxytocin receptor mRNA was determined by sucrose density gradient fractionation of bovine endometrial poly(A)+RNA and expression in frog oocytes to roughly 2000 base pairs (Morley et al. 1988). This has the capacity to encode the 57.5- to 65-kDa protein but presumably not the 78-kDa protein, arguing again for receptor heterogeneity.

9 Receptor Regulation

Growing evidence indicates that a finely tuned balance of oxytocin levels and receptor numbers accomodates the physiological requirement of tissue or organ sensitivity. For example, in the ewe and the cow, the amounts of tritiated oxytocin bound by plasma membranes from the endometrium and myometrium vary considerably during the estrous cycle, being low during the luteal phase, increasing several days before estrus to a maximum at estrus, and declining thereafter (Sheldrick and Flint 1985; Roberts et al. 1976; Soloff and Fields 1989). Oxytocin-mediated prostaglandin release also peaks at estrus, coinciding with the uterotonic response (Roberts et al. 1976). These data imply that the uterine receptor is an important determinant in mediating uterine sensitivity to oxytocin and that this is correlated with changes in receptor numbers. This is particularly intriguing because uterine sensitivity is inversely correlated during the cycle with circulating oxytocin levels. These are highest during the midluteal phase and correspond to those of the corpus luteum. The increase during ovulation to a peak in the midluteal phase (Schams et al. 1985; Wathes et al. 1984) probably meets the requirements of the ovary with respect to the life span of the corpus luteum (reviewed in Wathes 1984). Recent evidence indicates that the rise in the uterine oxytocin receptor number which is caused by 17β-estradiol (Hixon and Flint 1987; Jacobson et al. 1987) and which can be downregulated by progesterone via nuclear estrogen receptors (Takeda and Leavitt 1986; Leavitt et al. 1985) is correlated to uterine activity (Windmoller et al. 1983). The regulation of oxytocin receptors by sex steroids is of particular significance since the latter may form control loops by acting on the central nervous system to accommodate sexual and maternal behavior and physiological requirements. In fact, estradiol increases oxytocin receptor density in the ventromedial nucleus of the hypothalamus (de Kloet et al. 1986), a brain region controlling aspects of feeding, aggression, and sexual behavior (Nance 1976; Colpaert and Wiepkema 1976; Pfaff 1983).

References

Akaishi T, Sakuma Y (1985) Estrogen excites oxytocinergic but not vasopressinergic cells in the paraventricular nucleus of female rat hypothalamus. Brain Res 335:302–305

Antoni FA (1986) Oxytocin receptors in rat adenohypophysis: evidence from radioligand binding studies. Endocrinology 119:2393–2395

Audigier S, Barberis C (1985) Pharmacological characterization of two specific binding sites for neurohypophyseal hormones in hippocampal synaptic plasma membranes of the rat. EMBO J 4: 1407–1412

Bargmann W (1966) Neurosecretion. Int Rev Cytol 19:183–201

Barr PJ (1991) Mammalian subtilisins: the long-sought dibasic processing endoproteases. Cell 66:1–3

Barr PJ, Mason OB, Landsberg KE, Wong PA, Kiefer MC, Brake AJ (1991) cDNA and gene structure for a human subtilisin-like protease with cleavage specifitiy for paired basic amino acid residues. DNA Cell Biol 10:319–328

Bradbury AF, Smyth DG (1991) Peptide amidation. TIBS 16:112–115

Bradbury AF, Finnie MDA, Smyth DG (1982) Mechanism of C-terminal amide formation by pituitary enzymes. Nature 298:686–688

Cantau B, Barjon NJ, Chicot D, Baskevitch PP, Jard S (1990) Oxytocin receptors from LLC-PK1 cells: expression in oocytes. Am J Physiol 258:F963–972

Carsten ME, Miller JD (1985) Ca^{2+} release by inositol trisphosphate from Ca^{2+} transporting microsomes derived from uterine sarcoplasmic reticulum. Biochem Biophys Res Commun 130:1027–1031

Chadio SE, Antoni FA (1989) Characterization of oxytocin receptors in rat adenohypophysis using a radioiodinated receptor antagonist peptide. J Endocrinol 122:465–470

Charpak S, Armstrong WE, Mühlethaler M. Dreifuss JJ (1984) Stimulatory action of oxytocin on neurones of the dorsal motor nucleus of the vagus nerve. Brain Res 300:83–89

Chauvet MT, Hurpet D, Chauvet J, Acher R (1983) Identification of human neurophysins: complete amino acid sequences of MSEL- and VLDV-neurophysins. Proc Natl Acad Sci USA 80:2839–2843

Clamigirand C, Creminan C, Fahy C, Boussetta H, Cohen P (1987) Partial purification and functional properties of an endopeptidase from bovine neurosecretory granules cleaving prooxytocin/neurophysin peptides at the basic amino acid doublet. Biochemistry 26: 6018–6023

Colpaert FC, Wiepkema PR (1976) Effect of ventromedial hypothalamic lesions on spontaneous intraspecies aggression in male rats. Behav Biol 16:117–125

Dale HH (1906) On some physiological actions of ergot. J Physiol (Lond) 34:163–206

Davis L, Banker GA, Steward O (1987) Selective dendritic transport of RNA in hippocampal neurons in culture. Nature 330:477–479

de Kloet ER, Voorhuis DAM, Boschma Y, Elands S (1986) Estradiol modulates density of putative 'oxytocin receptors' in discrete rat brain regions. Neuroendocrinology 44: 415–421

Dierickx K, Vandesande F (1979) Immunocytochemical demonstration of separate vaso-pressin-neurophysin and oxytocin-neurophysin neurons in the human hypothalamus. Cell Tissue Res 196:203–212

Dirks RW, Raap AK, van Minnen J, Vreugdenhil E, Smit AB, van der Ploeg M (1989) Detection of mRNA molecules coding for neuropeptide hormones of the pond snail Lymnaea stagnalis by radioactive and non-radioactive in situ hybridization: a model study for mRNA detection. J Histochem Cytochem 37:7–14

Di Scala-Guenot D, Strosser MT, Freund-Mercier MR, Richard P (1990) Characterization of oxytocin-binding sites in primary rat brain cell cultures. Brain Res 524:10–16

Egan JJ, Saltis J, Wek SA, Simpson IA, Londos C (1990) Insulin, oxytocin and vasopressin stimulate protein kinase C activity in adipocyte plasma membranes. Proc Natl Acad Sci USA 87:1052–1056

el Alj A, Bonoris E, Cynoher E, Germain G (1990) Heterogeneity of oxytocin receptors in the pregnant rat myometrium near parturition. Eur J Pharmacol 186:231–238

Elands J, Resnik A, de Kloet ER (1990) Neurohypophyseal hormone receptors in the rat thymus, spleen, and lymphocytes. Endocrinology 126:2703–2710

Elands J, Beetsma A, Barberis C, De Kloet ER (1988) Topography of the oxytocin receptor system in rat brain: an autoradiographical study with a selective radioiodinated oxytocin antagonist. J Chem Neuroanatom 1:293–302

Evans JJ, Catt KJ (1989) Gonadotrophin-releasing activity of neurohypophyseal hormones: II. the pituitary oxytocin receptor mediating gonadotrophin release differs from that of corticotrophs. J Endocrinol 122:107–116

Fahrenholz F, Hackenberg M, Müller M (1988) Identification of a myometrial oxytocin re-ceptor protein. Eur J Biochem 174:81–85

Forsling ML (1986) Regulation of oxytocin release. In: Current Topics in Neuroendocri-nology 6. Neurobiology of Oxytocin. Springer, Berlin Heidelberg New York, pp 19–54

Forsling ML, Brimble MJ (1985) The role of oxytocin in salt and water balance. In: Oxyto-cin: clinical and laboratory studies. Elsevier, Amsterdam, pp 167–175

Fricker LD, Snyder SH (1983) Purification and characterization of enkephalin convertase, an enkephalin synthesizing carboxypeptidase. Eur J Biol Chem 258:10950–10955

Fricker LD, Evans CJ, Esch FS, Herbert E (1986) Cloning and sequence analysis of cDNA for bovine carboxypeptidase E. Nature 323:461–464

Fuller RS, Brake AJ, Thorner J (1989) Intracellular targeting and structural conservation of a prohormone-processing endoprotease. Science 246:482–486

Gainer H, Sarne Y, Brownstein MJ (1977a) Neurophysin biosynthesis: conversion of a pu-tative precursor during axonal transport. Science 195:1354–1356

Gainer H, Sarne Y, Brownstein MJ (1977b) Biosynthesis and axonal transport of rat neuro-hypophysial proteins and peptides. J Cell Biol 73:366–381

Gibson A, Shaw CR (1989) The effect of Ca-deprivation and of Ca-blocking drugs on oxytocin-induced contractions of the male mouse anococcygeus. J Pharm Pharmacol 41: 412–415

Giuditta A, Metafora S, Felsoni A, Del Rio A (1977) Factors for protein synthesis in the axoplasm of squid giant axons. J Neurochem 28:1393–1395

Giuditta A, Menichini E, Castigli E, Perrone Capano C (1990) Protein synthesis in the axo-nal territory. In: Neurology and Neurobiology 59. Stella AMG, de Vellis J, and Perez-Polo JR (eds) Regulation of Gene Expression in the Nervous System. Wiley, New York, pp 205–218

Giuditta A, Menichini E, Capano CP, Langella M, Martin R, Castigli E, Kaplan BB (1991) Active polysomes in the axoplasm of the squid giant axon. J Neurosci Res 28:18–28

Gordon-Weeks PR (1988) RNA transport in dendrites. Trends Neurosci 11:342–343

Hara Y, Battey J, Gainer H (1990) Structure of the mouse vasopressin and oxytocin genes. Mol Brain Res 8:319–324

Heierhorst J, Morley SD, Figueroa J, Krentler C, Lederis K, Richter D (1989) Vasotocin and isotocin precursors from the white sucker Catostomus commersoni: cloning and se-quence analysis of the cDNAs. Proc Natl Acad Sci USA 86:5242–5246

Hixon JE, Flint AP (1987) Effects of a luteolytic dose of oestradiol benzoate on uterine oxytocin receptor concentration, phosphoinositide turnover and prostaglandin $F_{2\alpha}$ secretion in sheep. J Repr Fertil 79:457–467

Ingram CD, Cutler KL, Wakerley JB (1990) Oxytocin excites neurones in the bed nucleus of the stria terminalis of the lactating rat in vitro. Brain Res 527:167–170

Ivell R (1986) Biosynthesis of oxytocin in the brain and peripheral organs. In: Neurobiology of Oxytocin. Springer, Berlin Heidelberg New York, pp 1–18 (Current topics in neuroendocrinology vol 6)

Ivell R, Richter D (1984) Structure and comparison of the oxytocin and vasopressin genes from rat. Proc Natl Acad Sci USA 81:2006–2010

Jacobson C, Riemer RK, Goldfien AC, Lykins D, Siiteri PK, Roberts JM (1987) Rabbit myometrial oxytocin and alpha2-adrenergic receptors are increased by estrogen but are differentially regulated by progesterone. Endocrinology 120:1184–1189

Jard S, Gaillard RC, Guillon G, Marie J, Schoenenberg P, Muller AF, Manning M, Sawyer WK (1986) Vasopressin antagonists allow demonstration of a novel type of vasopressin receptor in the rat adenohypophysis. Mol Pharmacol 30:171–177

Jard S, Barberis C, Audigier S, Tribollet E (1987) Neurohypophyseal hormone receptor systems in brain and periphery. In: de Kloet ER, Wiegant VM, de Wied D (eds) Prog Brain Res 72:173–186

Johnson MH, Everitt BJ (1984) Essential reproduction, 2nd edn. pp 300–327 Blackwell Scientific Publications, Oxford

Koenig E (1979) Ribosomal RNA in Mauthner axon: implications for a protein synthesizing machinery in the myelinated axon. Brain Res 174:95–107

Kovacs GL (1986) Oxytocin and behaviour. In: Ganten D, Pfaff D (eds) Neurobiology of oxytocin. Springer, Berlin Heidelberg New York, pp 91–128

Kursheed A, Sanborn BM (1989) Changes in intracellular free calcium in isolated myometrial cells: role of extracellular and intracellular calcium and possible involvement of guanine nucleotide-sensitive proteins. Endocrinology 124:17–23

Lasek RJ, Brady ST (1981) The axon: a prototype for studying expressional cytoplasm. Cold Spring Harbor Symp Quant Biol 46:113–124

Leavitt WW, Okulicz WC, McCracken JA, Schramm W, Robidoux WT (1985) Rapid recovery of nuclear estrogen receptor and oxytocin receptor in the ovine uterus following progesterone withdrawal. J Steroid Biochem 22:687–691

Lederis K, Goren HJ, Hollenberg MD (1985) Oxcytocin: an insulin-like hormone. In: Amico LA, Robinson AG (eds) Oxytocin: clinical and laboratory studies. Elsevier, Amsterdam, pp 284–302

Marc S, Leiber D, Harbon S (1986) Carbachol and oxytocin stimulate generation of inositol phosphate in guinea pig myometrium. FEBS Lett 201:9–14

Marc S, Leiber D, Harbon S (1988) Fluoroaluminates mimic muscarinic- and oxytocin-receptor-mediaded generation of inositol phosphates and contraction in the intact guinea-pig myometrium. Biochem J 255:705–713

Meyerhof W, Morley S, Schwarz J, Richter D (1988) Receptors for neuropeptides are induced by exogenous poly(A)+RNA in oocytes from Xenopus laevis. Proc Natl Acad Sci USA 85:714–717

Miller FD, Ozimek G, Milner RJ, Bloom FE (1989) Regulation of neuronal oxytocin mRNA by ovarian steroids in the mature and developing hypothalamus. Proc Natl Acad Sci USA 86:2468–2472

Mohr E, Schmitz E (1991) Functional characterization of estrogen and glucocorticoid responsive elements in the rat oxytocin gene. Mol Brain Res 9:293–298

Mohr E, Schmitz E, Richter D (1988) A single rat genomic DNA fragment encodes both the oxytocin and vasopressin genes separated by 11 kilobases and oriented in opposite transcriptional directions. Biochimie 70:649–654

Mohr E, Fehr S, Richter D (1991) Axonal transport of neuropeptide encoding mRNAs within the hypothalamo-hypophyseal tract of rats. EMBO J 10:2419–2424

Morley SD, Meyerhof W, Schwarz J, Richter D (1988) Functional expression of the oxytocin receptor in Xenopus laevis oocytes primed with mRNA from bovine endometrium. J Mol Endocrinol 1:77–81

Morley SD, Schönrock C, Heierhorst J, Figueroa J, Lederis K, Richter D (1990) Vasotocin genes in the teleost fish Catostomus commersoni: gene structure, exon-intron boundery, and hormone precursor organization. Biochemistry 29:2506–2511

Mühlethaler M, Sawyer WH, Manning MM, Dreifuss JJ (1983) Characterization of a uterine type oxytocin receptor in the rat hippocampus. Proc Natl Acad Sci USA 80: 6713–6717

Müller M, Soloff MS, Fahrenholz F (1989) Photoaffinity labelling of the oxytocin receptor in plasma membranes from rat mammary glands. FEBS Lett 242:333–336

Nance DM (1976) Sex difference in the hypothalamic regulation of feeding behaviour in the rat. Adv Psychol 3:75–123

Negoro H, Vissesuwan S, Holland RC (1973) Unit activity in the paraventricular nucleus of female rats at different stages of the reproductive cycle and after ovariectomy, with or without oestrogen or progesterone treatment. J Endocrinol 59:545–558

Nicholson HD, Swann RW, Burford GD, Wathes DC, Porter GD, Pickering BT (1984) Identification of oxytocin and vasopressin in the testis and in adrenal tissue. Regul Pept 8:141–146

Nörenberg U, Richter D (1988) Processing of oxytocin precursor: isolation of an exopeptidase from neurosecretory granules of bovine pituitaries. Biochim Biophys Res Comm 156:898–904

Okabe T, Goren JH, Lederis K, Hollenberg MD (1985) Oxytocin and glucose oxidation in the rat uterus. Regul Pept 10:269–279

Okazaki ME, Sagawa N, Okita JR, Bleasdale JE, MacDonald PC, Johnston JM (1981) Diacylglycerol metabolism and arachidonic acid release in human fetal membranes and decidua cells. J Biol Chem 256:7316–7321

Olson BR, Drutarosky MD, Stricker EM, Verbalis JG (1991) Brain oxytocin receptor antagonism blunts the effect of anorexigenic treatments in rats: evidence for central oxytocin inhibition of food intake. Endocrinology 129:785–791

Ott I, Scott JC (1910) The action of infundibulin upon the mammary secretion. Proc Soc Exp Biol Med 8:48–49

Pfaff DW (1983) Impact of estrogens on hypothalamic nerve cells: ultrastructural, chemical and electrical effects. Recent Prog Horm Res 39:127–179

Popescu LM, Nutu O, Panoiu C (1985) Oxytocin contracts the human uterus at term by inhibiting the myometrial Ca^{2+} extrusion pump. Biosci Rep 5:21–28

Raggenbass M, Tribollet E, Dubois-Dauphin M, Dreifuss JJ (1989) Correlation between oxytocin neuronal sensitivity and oxytocin receptor binding: an electrophysiological and autoradiographical study comparing rat and guinea pig hippocampus. Proc Natl Acad Sci USA 86:750–754

Rehbein M, Hillers M, Mohr E, Ivell R, Morley S, Schmale H, Richter D (1986) The neurohypophyseal hormones vasopressin and oxytocin: precursor structure, synthesis, and regulation. Biol Chem 367:695–704

Rhodes CH, Morell JI, Pfaff DW (1981) Distribution of estrogen-concentrating, neurophysin-containing magnocellular neurons in the rat hypothalamus as demonstrated by a technique combining steroid autoradiography and immunohistology in the same tissue. Neuroendocrinology 33:18–23

Richter D, Meyerhof W, Buck F, Morley SD (1991) Molecular biology of receptors for neuropeptide receptors. Curr Top Pathol 83:117–139

Rivera J, Lopez Bernal A, Varney M, Watson SP (1990) Inositol 1,4,5-trisphosphate and oxytocin binding in human myometrium. Endocrinology 127:155–162

Roberts JS, McCracken JA, Gavagan JE, Soloff Ms (1976) Oxytocin-stimulated release of prostaglandin $F_{2\alpha}$ from ovine endometrium in vitro: correlation with estrous cycle and oxytocin-receptor binding. Endocrinonolgy 99:1107–1114

Ruppert S, Scherer G, Schütz G (1984) Recent gene conversion involving bovine vasopressin and oxytocin precursor genes suggested by nucleotide sequence. Nature 308:554–557

Sausville E, Carney D, Battey J (1985) The human vasopressin gene is linked to the oxytocin gene and is selectively expressed in a cultured lung cancer cell line. J Biol Chem 260:10236–10241

Sawyer WH, Grzonka Z, Manning M (1981) Design of tissue specific agonists and antagonists. Mol Cell Endocrinol 22:117–134

Schams D, Schallenberg E, Meyer HHD, Bullermann B, Breitinger HJ, Euzenhofer G, Koll R, Kruip TAM, Walter DC, Karg H (1985) Ovarian oxytocin during the estrous cycle in cattle. In: Amico JA, Robinson AG (eds) Oxytocin, clinical and laboratory studies. Elsevier, Amsterdam, pp 317–334

Scharrer E, Scharrer B (1940) Secretory cells within the hypothalamus. Res Publ Assoc Res Nerv Ment Dis 20:170–194

Schmale H, Heinsohn S, Richter D (1983) Structural organization of the rat gene for the arginine vasopressin-neurophysin precursor. EMBO J 2:763–767

Schmidt A, Dreifuss JJ, Tribollet E (1991) Oxytocin receptors in rat kidney during development. Am J Physiol 259:872–881

Schrey MP, Read AM, Steer PJ (1986) Oxytocin and vasopressin stimulate inositol phosphate production in human gestational myometrium and decidua cells. Biosci Rep 6:613–619

Sheldrick EL, Flint AP (1985) Endocrine control of uterine oxytocin receptors in the ewe. J Endocrinol 106:249–258

Soloff MS, Fernstrom MA (1987) Solubilization and properties of oxytocin receptors from rat mammary gland. Endocrinology 120:2474–2482

Soloff MS, Fields MJ (1989) Changes in uterine oxytocin receptor concentrations throughout the estrous cycle of the cow. Biol Reprod 40:283–287

Soloff MS, Sweet P (1982) Oxytocin inhibition of $(Ca^{2+} + Mg^{2+})$ATPase activity in rat myometrial plasma membranes. J Biol Chem 257:10687–10693

Soloff MS, Beauregard G, Potier M (1988) Determination of the functional size of oxytocin receptor from mammary gland and uterine myometrium of the rat by radiation analysis. Endocrinology 122:1769–1772

Stassen FL, Heckman G, Schmidt D, Papadopoulos MT, Nambi P, Sarau H, Aiyar N, Gellai M, Kinter L (1988) Oxytocin induces a transient increase in cytosolic free [Ca2+] in renal tubular epithelial cells: evidence for oxytocin receptors on LLC-PK1 cells. Mol Pharmacol 33:218–224

Stoeckel ME, Freund-Mercier MJ, Palacios JM, Richard P, Porte A (1987) Autoradiographic localization of binding sites for oxytocin and vasopressin in the rat kidney. J Endocrinol 113:179–182

Swanson LW, Sawchenko PE (1983) Hypothalamic integration: organization of the paraventricular and supraoptic nuclei. Ann Rev Neurosci 6:269–324

Takeda A, Leavitt WW (1986) Temporal effects of progesterone on estrogen and oxytocin receptor in hamster uterus. J Steroid Biochem 25:219–224

Teitelbaum I (1991) Vasopressin-stimulated phosphoinositide hydrolysis in cultured rat inner medullary collecting duct cells is mediated by the oxytocin receptor. J Clin Invest 87:2122–2126

Tence M, Guillon G, Bottari S, Jard S (1990) Labelling of vasopressin and oxytocin receptors from the human uterus. Eur J Pharmacol 191:427–436

Tribollet E, Barberis C, Jard S, Dubois-Dauphin M, Dreifuss JJ (1988) Localization and pharmacological characterization of high affinity binding sites for vasopressin and oxytocin in the rat brain by light microscopic autoradiography. Brain Res 442:105–118

Tucker RP, Garner CC, Matus A (1989) In situ localization of microtubule-associated protein mRNA in the developing and adult rat brain. Neuron 2:1245–1256

Van Minnen J, Schallig HDFH (1990) Demonstration of insulin-related substances in the central nervous system of pulmonates and Aplysia californica. Cell Tissue Res 260:381–386

Van Tol HHM, Voorhuis DTAM, Burbach JPH (1987) Oxytocin gene expression in discrete hypothalamic magnocellular cell groups is stimulated by prolonged salt loading. Endocrinology 120:71–76

Van Tol HHM, Bolwerk ELM, Lin B, Burbach JPH (1988) Oxytocin and vasopressin gene expression in the hypothalamo-neurohypophyseal system of the rat during the estrous cycle, pregnancy, and lactation. Endocrinology 122:945–951

Verbalis JG, Mangione MP, Stricker EM (1991) Oxytocin produces natriuresis in rats at physiological plasma concentrations. Endocrinology 128:1317–1322

Walter P, Blobel G (1981) Translocation of proteins across the endoplasmic reticulum. II. Signal recognition protein (SRP) mediates the selective binding to microsomal membranes of in vitro assemble polysomes synthesizing secretory protein. J Cell Biol 91: 551–556

Wathes DC (1984) Possible actions of gonadal oxytocin and vasopressin. J Reprod Fertil 71:315–345

Wathes DC, Swann RW, Pickering BT (1984) Variations in oxytocin, vasopressin and neurophysin concentrations in the bovine ovary during the estrous cycle and pregnancy. J Reprod Fertil 71:551–557

Wilson T, Liggins GC, Whittaker DK (1988) Oxytocin stimulates the release of arachidonic acid and prostaglandin F2alpha from human decidual cells. Prostaglandins 35:771–780

Windmoller R, Lye SJ, Challis JR (1983) Estradiol modulation of ovine uterine activity. Can J Physiol Pharmacol 61:722–728

Wise RJ, Barr PB, Wong PA, Kiefer MC, Brake AJ, Kaufmann RJ (1990) Expression of a human proprotein processing enzyme: correct cleavage of the von Willebrand factor precursor at a paired basic amino acid site. Proc Natl Acid Sci USA 87:9378–9382

Yamaguchi K, Akaishi T, Negoro H (1979) Effect of estrogen treatment on plasma oxytocin and vasopressin in ovariectomized rats. Endocrinol Jpn 26:197–205

Yoshinaga K, Hawkins RA, Stocker JF (1969) Estrogen secretion by the rat ovary in vivo during the estrous cycle and pregnancy. Endocrinology 85:103–112

Young WS III, Reynolds K, Shepard EA, Gainer H, Castel M (1990) Cell-specific expression of the rat oxytocin gene in transgenic mice. J Neuroendocrinol 2:917–925

Zhao X, Gorewit RC (1987) Inositol-phosphate response to oxytocin stimulation in dispersed bovine mammary cells. Neuropeptides 10:227–233

Zingg HH, Lefebvre DL (1988) Oxytocin and vasopressin gene expression during gestation and lactation. Mol Brain Res 4:1–6

Rev. Physiol. Biochem. Pharmacol., Vol. 121
© Springer-Verlag 1992

Peptidergic Sensory Neurons in the Control of Vascular Functions: Mechanisms and Significance in the Cutaneous and Splanchnic Vascular Beds

PETER HOLZER

Contents

University of Graz, Department of Experimental and Clinical Pharmacology,
Universitätsplatz 4, A-8010 Graz, Austria

Abbreviations: CGRP, calcitonin gene-related peptide; EDRF, endothelium-derived relaxing factor; NANC, nonadrenergic noncholinergic; NKA, neurokinin A,; NKB, neurokinin B; NO, nitric oxide; NPY, neuropeptide Y; SP, substance P; VIP, vasoactive intestinal polypeptide

1 Introduction

Tissue blood flow is regulated by local autoregulatory mechanisms, factors released from endothelial cells, hormones and neural factors. The role of autonomic neurons, notably of sympathetic vasoconstrictor neurons, is well studied. However, even after blockade of cholinergic and noradrenergic transmission, blood vessels can actively be dilated in response to nerve stimulation. The nonadrenergic noncholinergic (NANC) vasodilator neurons have long eluded identification, but there is now good evidence that NANC vasodilatation in many vascular beds is mediated by nerve fibres that utilize peptides such as substance P (SP) and calcitonin gene-related peptide (CGRP) as their transmitters. The perivascular NANC neurons, however, are not identical with classical autonomic neurons whose role is to control vascular effector functions – they are in fact sensory neurons ca-

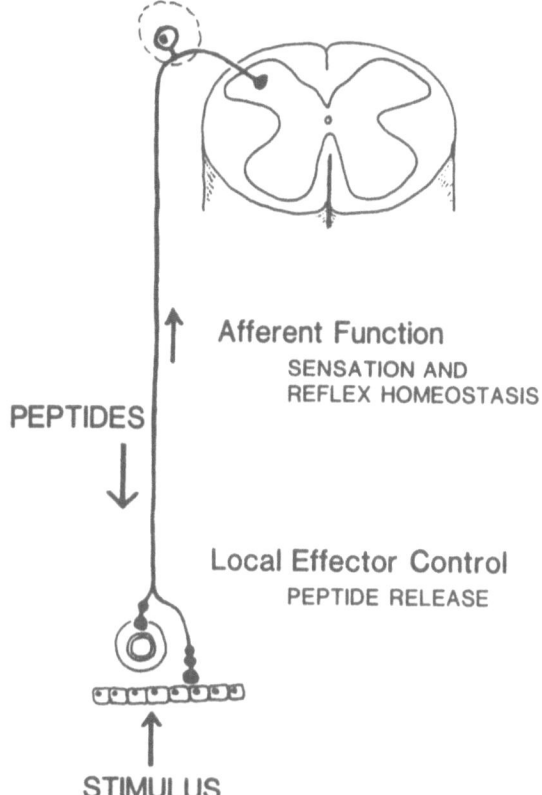

Fig. 1. Dual role of peptidergic afferent neurons. These neurons are sensory neurons with unmyelinated or thinly myelinated axons. The information they perceive is transmitted to the central nervous system (spinal cord or brainstem) to produce sensation and initiate autonomic homeostatic reflexes. In addition, peptide transmitters can be released from their peripheral endings to control the function of adjacent effector systems, e.g. blood vessels. Peptidergic afferent neurons are well equipped for this peripheral function because the bulk of peptides synthesized in the neuronal somata is transported towards the peripheral endings (Brimijoin et al. 1980; Keen et al. 1982)

pable of monitoring their chemical and physical environment and primary afferent neurons that convey this information to the central nervous system. Recognition of this dual role of perivascular peptidergic afferent neurons (Fig. 1) has lit up completely new aspects in the neural control of circulation. It has also stirred reconsideration of the principal organization of the autonomic nervous system (see Prechtl and Powley 1990).

Although the significance and mode of action of afferent neurons in the control of vascular functions is only now being appreciated, the discovery of primary afferent vasodilator fibres dates back to the last century. Stricker (1876) was the first to observe that stimulation of the peripheral ends of cut dorsal roots induced vasodilatation in the skin area supplied by the transected afferent nerve fibres. As efferent fibres are absent from the dorsal roots (see Coggeshall 1980) and removal of the dorsal root ganglia abolished the vasodilator response (Bayliss 1901), it was inferred that cutaneous vasodilatation resulted from antidromic conduction of nerve impulses in afferent nerve fibres (Bayliss 1901). In the rat, this "antidromic vasodilatation" is associated with an increase in vascular permeability leading to extravasation of plasma proteins and formation of oedema (N. Jancsó et al. 1967). Hyperaemia and increased vascular permeability are key traits of inflammation, and "neurogenic inflammation" is a term which was first used to denote the inflammatory response of the skin to application of the irritant mustard oil, a response that depends on the afferent innervation of the tissue (Bruce 1910; N. Jancsó et al. 1967).

Peptidergic afferent neurons regulate vascular functions in many somatic and visceral tissues (see Lundberg and Saria 1987; Barnes et al. 1988; Chahl 1988; Holzer 1988; Lembeck and Holzbauer 1988; Maggi and Meli 1988). Two mechanisms of action can be differentiated. On the one hand, vasoactive peptides released from the peripheral nerve terminals exert a local control over vascular effector functions. On the other hand, afferent neurons also participate in the reflex control of circulation via the autonomic nervous system. The present article highlights these implications of afferent neurons in two model systems, the cutaneous and splanchnic circulation. These two systems have been chosen not only because they have been studied extensively, but also because they enable us to point out the diversity of vascular control mechanisms governed by peptidergic afferent neurons.

2 Nature of Afferent Neurons Regulating Vascular Functions

2.1 Capsaicin-Sensitive Neurons

The afferent neurons that are involved in local blood flow regulation are sensitive to capsaicin (Fig. 2, Table 1), a property which has greatly facilitated their anatomical, neurochemical and functional investigation (N. Jancsó et al. 1967). Low doses of capsaicin (in the microgram per kilogram range) induce transient excitation of thin primary afferent neurons whilst systemic administration of high doses of the drug (in the milligram per kilogram range) to small rodents causes long-lasting damage of these neurons (see Buck and Burks 1986; G. Jancsó et al. 1987; Szolcsányi 1990a; Holzer 1991). The extent of injury (ultrastructural changes or degeneration of C fibre and Aδ fibre afferent neurons) depends on the dosage, route of administration, animal species and age of the animals (see G. Jancsó et al.

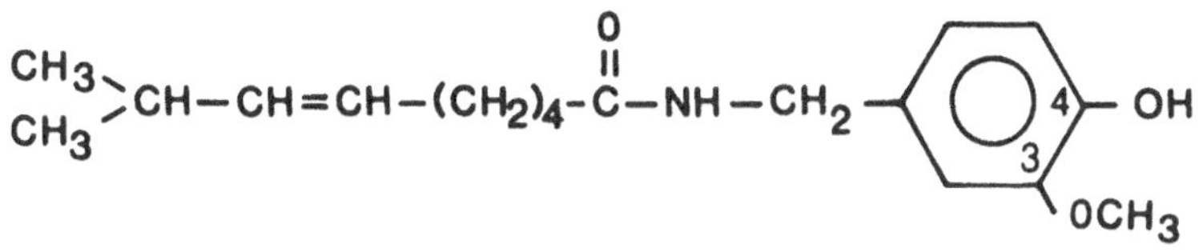

$$CH_3 > CH-CH=CH-(CH_2)_4-\overset{\overset{\displaystyle O}{\|}}{C}-NH-CH_2- \underset{3}{\bigcirc}_4- OH$$
$$OCH_3$$

Fig. 2. Chemical structure of capsaicin

Table 1. Brief summary of capsaicin's main sensory neuron-selective actions (Holzer 1991) which can be utilized in the functional investigation of peptidergic afferent neurons

Acutely stimulant effect:	Excitation
	Nociception
	Peptide release
Long-term neurotoxic effect:	Ultrastructural changes
	Degeneration
	Defunctionalization
	Peptide depletion

1987; Szolcsányi 1990a; Holzer 1991). Half-maximal excitation of isolated or cultured rat afferent neurons is produced by 0.1–0.5 μmol l^{-1} capsaicin, whereas concentrations above 1 μmol l^{-1} exert a definitely neurotoxic action within 5–20 min (see Holzer 1991). These actions of capsaicin (Table 1) are mediated by specific receptors coupled to non-selective cation channels in the cell membrane, and intracellular accumulation of Ca^{2+} and NaCl are the major factors that determine capsaicin's neurotoxicity (see Bevan and Szolcsányi 1990; Holzer 1991).

The "sensory neuron-selective" effects of capsaicin (Table 1) are confined to mammals and show marked species differences even within this class of vertebrates (Holzer 1991). However, its specificity for afferent neurons is not absolute, and some of the acute effects of the drug on blood vessels appear to result from a direct action on vascular muscle (Donnerer and Lembeck 1982; Duckles 1986; Bény et al. 1989; Edvinsson et al. 1990; Moritoki et al. 1990) and endothelium (Kenins et al. 1984). The label "capsaicin sensitive" is applied to those afferent neurons which acutely are *excited* and in the long term are *damaged* by the drug. This population of neurons is heterogeneous and comprises most but not all primary afferent neurons with small cell bodies and unmyelinated (C fibre) axons and some afferent neurons with thinly myelinated (Aδ fibre) axons. With regard to sensory modality, capsaicin-sensitive afferent neurons encompass chemoceptors, chemonociceptors, polymodal nociceptors and some warmth receptors (see Szolcsányi 1990a; Holzer 1991).

2.2 Perivascular Peptidergic Neurons

Capsaicin-sensitive and fine afferent neurons in general contain a number of bioactive peptides (Table 2) including CGRP, the tachykinins SP and neurokinin A (NKA), vasoactive intestinal polypeptide (VIP), somatostatin and dynorphin, which are localized within synaptic vesicles. Apart from regional and species differences (Bucsics et al. 1983; J. Donnerer, personal communication), there are pathway-specific patterns of the co-existence of these peptides in afferent neurons (Gibbins et al. 1987a, b; O'Brien et al. 1989), and it is likely that heterogeneity in "chemical coding" reflects functional heterogeneity (Costa et al. 1986; Mayer and Baldi 1991). Criteria to firmly establish peptides as transmitters of afferent neurons have thus far only been met for SP and CGRP. After synthesis of afferent neurons in the somata (Keen et al. 1982; Rosenfeld et al. 1983; Nakanishi 1987), the bulk of peptide is transported to the peripheral endings of afferent neurons (Brimijoin et al. 1980; Keen et al. 1982). Stimulation causes release of SP, NKA and CGRP from the peripheral endings of afferent neurons, and there is ample evidence that the co-released peptides interact in the control of vascular functions.

Table 2. Some bioactive peptides in capsaicin-sensitive primary afferent neurons

Peptide	Selected references
Bombesin/gastrin-releasing peptide	Decker et al. 1985
Calcitonin gene-related peptide	Rosenfeld et al. 1983; Gibbins et al. 1985, 1987a; Skofitsch and Jacobowitz 1985b; Wharton et al. 1986; Franco-Cereceda et al. 1987b
Cholecystokinin	G. Jancsó et al. 1981; Gibbins et al. 1987a
Corticotropin-releasing factor	Skofitsch et al. 1985b
Dynorphin	Gibbins et al. 1987a; Weihe 1990
Galanin	Skofitsch and Jacobowitz 1985a
Leucine enkephalin	Weihe 1990
Neurokinin A	Maggio and Hunter 1984; Hua et al. 1985
Peptide histidine methionine	Chéry-Croze et al. 1989
Somatostatin	G. Jancsó et al. 1981; Gamse et al. 1981; Nagy et al. 1981
Substance P	Jessell et al. 1978; Gamse et al. 1980; Nagy et al. 1980; G. Jancsó et al. 1981
Vasoactive intestinal polypeptide	G. Jancsó et al. 1981; Skofitsch et al. 1985b

SP-containing varicose and smooth nerve fibres of dorsal root ganglion origin innervate blood vessels throughout the body, but the density of fibres around arteries is in general higher than around veins (Furness et al. 1982; Barja et al. 1983; Gibbins et al. 1988). In the guinea pig, the para- and perivascular network of afferent SP-containing fibres varies from tissue to tissue, being especially dense in large arteries close to the heart (Furness et al. 1982; Papka et al. 1984). The density of innervation decreases as more peripheral arterial beds are approached, except that the mesenteric arteries receive a particularly rich supply (Furness et al. 1982; Barja et al. 1983). The veins of peripheral vascular beds contain only few SP-immunoreactive nerve fibres. SP-containing axons are primarily found in the connective tissue (tunica adventitia) surrounding the vessels and at the border between adventitia and media (muscle layer) (Furness et al. 1982; Barja et al. 1983). Their predominant orientation in many vessels is circumferential although, e.g. in the mesenteric and cerebral arteries, an interlacing network of fibres prevails (Furness et al. 1982; Barja et al. 1983).

The perivascular distribution of afferent nerve fibres containing CGRP is very similar to that of SP-containing fibres (Gibbins et al. 1985, 1987a, b; Lundberg et al. 1985; Mulderry et al. 1985; Uddman et al. 1986; Franco-Cereceda et al. 1987b; Ishida-Yamamoto et al. 1989; Kawasaki et al. 1990a). Small arteries in the limbs and in the respiratory, gastrointestinal and urogenital tracts of rat (Gibbins et al. 1985; Muldery et al. 1985; Wharton et al. 1986; Sternini et al. 1987; Su et al. 1987; Green and Dockray 1988; Wimalawansa and MacIntyre 1988; Del Bianco et al. 1991) and guinea pig (Uddman et al. 1986; Wharton et al. 1986) receive a particularly rich innervation by CGRP-immunoreactive nerve fibres which occur both in the adventitia and media. SP and CGRP co-exist in the perivascular afferent axons, but the overlap is not complete inasmuch as more fibres contain CGRP than SP (Gibbins et al. 1985, 1987a; Y. Lee et al. 1985; Matsuyama et al. 1986; Uddman et al. 1986; Wanaka et al. 1986; Ju et al. 1987; Su et al. 1987; Galligan et al. 1988; Green and Dockray 1988; S.M. Louis et al. 1989a; O'Brien et al. 1989; Helke and Niederer 1990; Maynard et al. 1990).

It should not go unnoticed here that SP is also present in endothelial cells (Loesch and Burnstock 1988; Linnick and Moskowitz 1989; Ralevic et al. 1990). This endothelial SP system is not affected by capsaicin pretreatment and seems to be activated by increased flow through the vascular bed (Ralevic et al. 1990). The precise functional significance of this system and any possible relationship to afferent nerve endings is not yet known.

2.3 Pharmacology of Tachykinin and CGRP Receptors

2.3.1 Receptor Classification and Agonists

The transmitters released from afferent nerve endings exert their actions on the vascular and other systems by interaction with specific receptors on the effector cells. As regards tachykinins, three distinct types of receptors have been identified by functional and binding studies: NK_1, NK_2 and NK_3 receptors (Regoli et al. 1989; Guard and Watson 1991). Their existence has been confirmed by the identification and cloning of three distinct cDNA sequences (Ohkubo and Nakanishi 1991). All three tachykinin receptors belong to the family of G protein-coupled receptors and are linked to the phosphoinositide transmembrane signalling pathway (Guard and Watson 1991). The proposed presence of an additional type of tachykinin receptor, the NK_4 receptor, remains to be proved (Guard and Watson 1991).

Pharmacologically, NK_1, NK_2 and NK_3 receptors are defined on the basis of different agonist and antagonist affinities. The three principal tachykinins occurring in mammals, SP, NKA (also called substance K) and neurokinin B (NKB; also called neuromedin K) show little selectivity towards the different tachykinin receptors and are capable of interacting with all receptor types. However, there are synthetic tachykinin analogues that display considerable receptor selectivity. For instance, SP methyl ester and $[Sar^9,Met(O_2)^{11}]$-SP are selective NK_1 receptor agonists, with negligible activity at NK_2 and NK_3 receptors (Regoli et al. 1989; Guard and Watson 1991). $[\beta\text{-Ala}^8]$-NKA_{4-10} (Rovero et al. 1989) is a selective NK_2 receptor agonist, while succinyl-$[Asp^6,MePhe^8]$-SP_{6-11} (SENKTIDE; Wormser et al. 1986) and $[MePhe^7]$-NKB (Regoli et al. 1989) are selective NK_3 receptor agonists.

Functional and binding studies have established that the biological actions of CGRP are also mediated by specific receptors for this peptide, but the molecular identification of these receptors has not yet been accomplished. Analysis of the structure–activity relationship of different CGRP analogues and fragments suggests the existence of at least two subtypes of CGRP receptors (Tippins et al. 1986; Maton et al. 1988, 1990; Dennis et al. 1989, 1990). It is not yet clear, however, whether the two molecular forms of CGRP occurring in humans and rat, CGRP-α (or CGRP-I) and CGRP-β (or CGRP-II) interact with the same or different CGRP receptor subtypes. The major transduction mechanism operated by CGRP receptors appears to be the adenylate cyclase/cyclic AMP system (Kubota et al. 1985; Hirata et al. 1988; Crossman et al. 1990).

2.3.2 Receptor Antagonists

The occurrence of three tachykinin receptor types has been confirmed by the development of competitive and selective receptor antagonists, some of which point to the existence of further tachykinin receptor subtypes. Replacement of certain L-amino acids in the sequence of SP by D-amino acids led to the discovery of a class of competitive antagonists exemplified by [D-Pro^2,D-Trp^{7,9}]-SP and [D-Arg^1,D-Trp^{7,9},Leu^{11}]-SP (SPANTIDE I). This first generation of tachykinin antagonists is characterized by low activity (pA$_2$ < 7), restricted selectivity for different tachykinin receptor types and non-specific effects including antagonism of bombesin, release of histamine and neurotoxic effects (Håkanson and Sundler 1985).

Further modification of the tachykinin amino acid sequence gave way to competitive tachykinin antagonists characterized by improved potency (pA$_2$ > 7), selectivity towards NK$_1$ and NK$_2$ receptors, and lack of non-specific effects. For instance, [D-Pro^9(spiro-γ-lactam)Leu^{10},Trp^{11}]-physalaemin (GR-82334) is a metabolically stable antagonist at NK$_1$ receptors (Hagan et al. 1991). Other compounds including the linear peptide H-Asp-Tyr-D-Trp-Val-D-Trp-D-Trp-Arg-NH$_2$ (MEN-10207; Maggi et al. 1990), the cyclic hexapeptide cyclo[Gln-Trp-Phe-Gly-Leu-Met] (L-659877; McKnight et al. 1991) and the pseudopeptide analogue [Leu^9ψ-(CH$_2$NH)Leu^{10}]-NKA$_{4\text{-}10}$ (MDL-28564; Buck et al. 1990) have been found to selectively antagonize NK$_2$ receptors. Because these NK$_2$ receptor-selective antagonists are differently active in different NK$_2$ receptor assays, a heterogeneity of NK$_2$ receptors has been proposed (Buck et al. 1990; Maggi et al. 1990; Patacchini et al. 1991). NK$_3$ receptor antagonists of similar selectivity are not yet available (Drapeau et al. 1990).

A new era in tachykinin receptor pharmacology began with the discovery of highly active (pA$_2$ > 8) non-peptide tachykinin antagonists. Two of these compounds, (2S,3S)-cis-2-(diphenylmethyl)-N-[(2-methoxyphenyl)-methyl]-1-azabicyclo[2.2.2]octan-3-amine (CP-96345; Snider et al. 1991) and (3aR,7aR)-7,7-diphenyl-2-[1-imino-2-(2-methoxyphenyl)-ethyl]perhydroisoindol-4-one (RP-67580; Garrett et al. 1991), are selective NK$_1$ receptor antagonists. The activity of CP-96345 is species dependent inasmuch as this antagonist is 30-120-fold less active in the rat and mouse than in other mammalian species including humans (Beresford et al. 1991). Possible limitations of the usefulness of CP-96345 in terms of non-specific effects (Constantine et al. 1991; Lembeck et al. 1992) remain to be sorted out. Another non-peptide compound, (-)-N-methyl-N[4-acetylamino-4-phenylpiperidino-2-(3,4-dichlorophenyl)butyl]benzamide (SR-48968) has been reported to be a potent and selective antagonist at NK$_2$ receptors (Emonds-Alt et al. 1992).

Modification of the amino-acid sequence of CGRP-α has led to the discovery of competitive and specific CGRP receptor antagonists (Chiba et al. 1989; Dennis et al. 1989, 1990; Maton et al. 1990; Mimeault et al. 1991). Human $CGRP_{8-37}$ is the most potent ($pA_2 \sim 7$) CGRP antagonist available to date and appears to differentiate between two CGRP receptor subtypes called $CGRP_1$ and $CGRP_2$ receptors. $CGRP_1$ receptors are characterized by their sensitivity to the antagonistic actions of $CGRP_{8-37}$, whereas $CGRP_2$ receptors are resistant to $CGRP_{8-37}$ (Dennis et al. 1990; Mimeault et al. 1991).

2.3.3 Termination of Agonist-Receptor Interaction by Peptidases

Diffusion and degradation by peptidase enzymes seem to be the main mechanisms by which peptides released from nerve terminals are inactivated and their action is terminated. Several peptidases capable of degrading tachykinins have been described, including prolyl endopeptidase (post-proline cleaving enzyme, EC 3.4.21.26; S. Blumberg et al. 1980), a membrane-bound SP-degrading enzyme (C.M. Lee et al. 1981), angiotensin converting enzyme (peptidyl dipeptidase A or kininase II, EC 3.4.15.1; Yokosawa et al. 1983) and endopeptidase 24.11 (neutral endopeptidase or enkephalinase, EC 3.4.24.11; Matsas et al. 1983). Endopeptidase 24.11 is a membrane-bound enzyme present in a variety of tissues including the vascular system, skin and gastrointestinal tract (Bunnett et al. 1985; M.E. Hall et al. 1989; Rouissi et al. 1990; Nadel 1992). Several findings underline the importance of this enzyme for the termination of the action of tachykinins. Inhibitors of endopeptidase 24.11, such as phosphoramidon or thiorphan, enhance the amount of tachykinins present at the site of release (Geppetti et al. 1989; Martins et al. 1991), increase the potency of tachykinins and prolong the duration of their action (Sekizawa et al. 1987; Frossard et al. 1989; J.M. Hall et al. 1990; Rouissi et al. 1990; Nadel 1992). Conversely, administration of recombinant neutral endopeptidase attenuates SP-induced plasma protein extravasation in the guinea pig skin (Rubinstein et al. 1990).

Although the precise inactivation mechanisms for CGRP are less well studied than for the tachykinins, it appears as if the biological actions of this peptide are terminated by enzymic catabolism as well (Brain and Williams 1988, 1989). Peptidases are not peptide selective, and endopeptidase 24.11 is able to catabolize both tachykinins and CGRP (Le Grevès et al. 1989).

3 Afferent Neurons in Cutaneous Circulation

3.1 Innervation of the Cutaneous Vascular Bed
by Peptidergic Afferent Neurons

Blood vessels in the skin of cat, rat, guinea pig, and pig are innervated by nerve fibres containing CGRP, SP and other peptides (Hökfelt et al. 1975; Cuello et al. 1978; Gibbins et al. 1985, 1987a, 1988; Franco-Cereceda et al. 1987b; Dalsgaard 1988; Ishida-Yamamoto et al. 1989; S.M. Louis et al. 1989a; Weihe 1990; Alving et al. 1991b). Axons containing CGRP supply adventitia and media, and come close to endothelial cells (Ishida-Yamamoto et al. 1989). Capsaicin pretreatment ablates about 50% of the unmyelinated axons in the rat skin (Chung et al. 1990), diminishes the retrograde labelling of dorsal root ganglion cells by > 90% (S.M. Louis et al. 1989a) and depletes SP and CGRP from the skin (Gamse et al. 1980; Holzer et al. 1982; Franco-Cereceda et al. 1987b; Geppetti et al. 1988; Kashiba et al. 1990; Alving et al. 1991b), which indicates that primary afferent neurons contribute significantly to the peptidergic innervation of the skin. This contention is supported by retrograde tracing of cutaneous nerve fibres to peptide-containing cell bodies in the dorsal root ganglia of the rat (Molander et al. 1987; S.M. Louis et al. 1989a; Kashiba et al. 1991). Whereas CGRP is contained in both small and large somata, SP occurs in small and medium-sized cell bodies and somatostatin is confined to small somata of rat afferent neurons (Molander et al. 1987; McCarthy and Lawson 1989, 1990; Kashiba et al. 1991). Peptide-containing fibres also innervate the human skin (Dalsgaard et al. 1983; Hartschuh et al. 1983; Björklund et al. 1986; Franco-Cereceda et al. 1987b; Dalsgaard 1988), where CGRP co-exists with SP and somatostatin in axons innervating small arteries, arterioles and capillaries (Gibbins et al. 1987b; Wallengren et al. 1987).

3.2 Local Regulation of Cutaneous Circulation

3.2.1 Neuropeptide Release in the Skin

Peptides such as SP, NKA and CGRP are released from the peripheral endings of afferent neurons when they are stimulated (see Holzer 1988; Maggi 1991). Antidromic stimulation of the sciatic nerve (White and Helme 1985) or noxious heat applied to the rat skin (Helme et al. 1986; Yonehara et al. 1987, 1991) leads to release of SP and CGRP, but not NKA, into the subcutaneous space, an effect that is inhibited by chemical or surgical ablation of afferent nerve fibres (Yonehara et al. 1987, 1991).

Capsaicin is also able to release CGRP from the rat paw skin in vitro (Donnerer and Stein 1992) and the hamster cheek pouch in vivo (Raud et al. 1991), and a major proportion of circulating CGRP in the rat seems to originate from capsaicin-sensitive perivascular nerve fibres (Zaidi et al. 1985; Diez Guerra et al. 1988). In the isolated perfused rabbit ear, both capsaicin and thermal stimulation lead to release of SP (Amann et al. 1990), and scalding injury in the dog enhances the release of SP into the lymph draining the damaged tissue (Johnsson et al. 1986). Similarly, occlusion of the human arm increases the SP concentration in the venous blood (Henriksen et al. 1986). The concentrations of SP and VIP in the fluid of human skin blisters are elevated under conditions of inflammatory skin diseases (Wallengren et al. 1986) as are the VIP levels in venous plasma after transcutaneous nerve stimulation (Kaada et al. 1984). The origin of VIP, though, is not known, and it has not yet been examined whether the other peptides present in cutaneous afferent nerve fibres (Table 1) are released upon stimulation.

3.2.2 Neuropeptide Receptors on Cutaneous Blood Vessels

Autoradiography has shown that NK_1 tachykinin receptors are present on the endothelium of capillaries in the human skin (Deguchi et al. 1989), and on arterioles and postcapillary venules in the rat footpad skin, whereas NK_2 receptor sites are absent (O'Flynn et al. 1989). The highest concentrations of CGRP binding sites are found in peripheral arteries of the rat limb and mesentery (Wimalawansa and MacIntyre 1988). Further analysis has revealed that CGRP receptors occur both on vascular muscle and endothelium (Hirata et al. 1988; Gates et al. 1989).

3.2.3 Vasodilatation

3.2.3.1 Nature of Afferent Vasodilator Fibres

In contrast to sympathetic vasoconstriction which is short lasting, it is typical of antidromic vasodilatation that hyperaemia outlasts the period of nerve stimulation. Antidromic vasodilatation is most obvious when electrical stimulation is strong enough to recruit unmyelinated (C) fibres (Hinsey and Gasser 1930; Low and Westerman 1989; Koltzenburg et al. 1990) which are connected to nociceptors (Celander and Folkow 1953a; Blumberg and Wallin 1987; Magerl et al. 1987), but it has recently been shown that also some thin myelinated (Aδ) fibres can give rise to cutaneous hyperaemia (Lynn 1988; Jänig und Lisney 1989). In frogs only Aδ fibres are able to induce antidromic vasodilatation (Khayutin et al. 1991). The magnitude of hyperaemia depends on the number and frequency of the stimuli delivered to the nerve (Celander and Folkow 1953b; Magerl et al. 1987; Szolcsányi 1988; Jänig and Lisney 1989), yet only one or two impulses or a frequency of 0.025 Hz are sufficient to elicit some vasodilatation as detected by laser Doppler velocimetry (Magerl et al. 1987; Lynn 1988; Lynn and Shakhanbeh 1988a; Szolcsányi 1988; Scolcsányi et al. 1992). Most, if not all, of the vasodilator nerve fibres are sensitive to the excitatory and neurotoxic actions of capsaicin. Topical or intradermal administration of capsaicin to the skin of humans (N. Jancsó et al. 1968; Helme and McKernan 1985), pig (Franco-Cereceda and Lundberg 1989; Alving et al. 1991a, b) and rabbit (Buckley et al. 1990) causes hyperaemia arround the application site. Conversely, antidromic vasodilatation in the rat is blocked by capsaicin-induced ablation of afferent neurons (Lembeck and Holzer 1979; Lembeck and Donnerer 1981b; Gamse and Saria 1987; Scolcsányi 1988; Low and Westerman 1989; Scolcsányi et al. 1992). Topical defunctionalization of capsaicin-sensitive nerve terminals inhibits cutaneous hyperaemia evoked by intraneural stimulation of the peroneal nerve (Blumberg and Wallin 1987) or percutaneous stimulation of nociceptors (Magerl et al. 1987).

3.2.3.2 Vasodilator Activity of Sensory Neuropeptides

Substance P causes vasodilatation in many vascular beds (see B. Pernow 1983). Infused close arterially to the hind leg of the rat, SP increases venous outflow indicative of vasodilatation (Lembeck and Holzer 1979; Lembeck and Donnerer 1981a; Lembeck et al. 1982; Holzer-Petsche et al. 1985), and SP infused into the brachial artery of humans induces hyperaemia in the skin and muscle of the forearm (Eklund et al. 1977; Benjamin et al. 1987). This effect arises from dilatation of both arteries and veins (Benjamin et al. 1987). Blood flow in the femoral artery (Hallberg and Pernow 1975) and skin of the dog (Burcher et al. 1977) and rat (Jansen et al. 1989) is also augmented by intravenous administration of SP. Intradermal injection of SP enhances cutaneous blood flow in the guinea pig (Woodward et al. 1985) but fails to change it in the rat dorsal skin (Brain and Williams 1988, 1989). However, when perfused over a blister base in the rat footpad skin, SP causes hyperaemia (Andrews and Helme 1989), as does topical administration of SP to the hamster cheek pouch (Raud et al. 1991).

The hypotensive effect and direct dilator action of tachykinins on arteries and arterioles are mediated by NK_1 tachykinin receptors (Holzer-Petsche et al. 1985; Maggi et al. 1985; D'Orléans-Juste et al. 1986; Andrews and Helme 1989; Couture et al. 1989; Constantine et al. 1991; Lembeck et al. 1992) and depend on the formation of endothelium-derived relaxing factor (EDRF) (Zawadzki et al. 1981; Furchgott 1984; Altura et al. 1985;

D'Orléans-Juste et al. 1985, 1986; Bolton and Clapp 1986; J. Pernow 1989; Stewart-Lee and Burnstock 1989; Whittle et al. 1989) and EDRF-induced formation of cyclic GMP in vascular smooth muscle (Schini et al. 1990). The major EDRF has now been identified as nitric oxide (NO) which is formed from L-arginine by the enzyme NO synthase (see Moncada et al. 1991). Mast cell-derived secondary mediators of the vascular effects of SP such as histamine are discussed in section 3.2.6.4.

CGRP is one of the most active vasodilator substances in many vascular beds including the skin, and, depending on the tissue under study, the two molecular forms of CGRP, CGRP-α and CGRP-β, are either equally (Franco-Cereceda et al. 1987a; G. Williams et al. 1988) or differently (Bauerfeind et al. 1989; Beglinger et al. 1991) potent in this respect. Administration of CGRP into the brachial artery of humans elicits arterial dilatation in the forearm (Benjamin et al. 1987; Thom et al. 1987; Ando et al. 1992), and intravenous infusion of CGRP-α increases blood flow in the human skin (Beglinger et al. 1991). Intradermal injection of CGRP (≥ 1 pmol) leads to a slowly developing but intense hyperaemia (Brain et al. 1985, 1986, 1990; Brain and Williams 1988, 1989; Pedersen-Bjergaard et al. 1991) which persists for hours (Piotrowski and Foreman 1986; Fuller et al. 1987; Wallengren and Håkanson 1987; G. Williams et al. 1988). The erythema is surrounded by an area of pallor (Piotrowski and Foreman 1986; Wallengren and Håkanson 1987). Blood flow in the rat skin is augmented by both intradermally (Brain and Williams 1988, 1989) and intravenously (Kjartansson et al. 1988; Jansen et al. 1989) injected CGRP. Vasodilatation is also seen after topical administration of CGRP to the hamster cheek pouch (Raud et al. 1991). The arteriolar dilatation evoked by this peptide seems to arise from a direct action on the vessels since it is not affected by histamine H_1 antagonists (Piotrowski and Foreman 1986; Fuller et al. 1987; Wallengren and Håkanson 1987), local anaesthetics (Wallengren and Håkanson 1987) and acetylsalicylic acid (Fuller et al. 1987).

The receptors mediating CGRP-induced hypotension (Donoso et al. 1990) and vasodilatation in certain vascular beds (Gardiner et al. 1990) have been classified as $CGRP_1$ receptors which, unlike $CGRP_2$ receptors, are antagonized by $CGRP_{8-37}$ (Dennis et al. 1990). The dilator effect of CGRP on arterioles in the rabbit skin is also inhibited by $CGRP_{8-37}$ (Hughes and Brain 1991). Endothelial NO does not seem to play a role in the CGRP-induced hyperaemia in the rat skin (Ralevic et al. 1992). CGRP-evoked dilatation of non-cutaneous arteries and arterioles is either endothelium dependent (Brain et al. 1985; Grace et al. 1987; Thom et al. 1987; Fiscus et al. 1991) or endothelium independent and mediated via stimulation of adenylate cyclase in vascular smooth muscle (Kubota et al. 1985;

Crossman et al. 1987, 1990; Greenberg et al. 1987; Shoji et al. 1987; Hirata et al. 1988; J. Pernow 1989; Maynard et al. 1990). Both relaxant mechanisms may be operated by CGRP in parallel (Bråtveit et al. 1991; Samuelson and Jernbeck 1991). In addition, CGRP leads to an EDRF-dependent increase in cyclic AMP levels in the rat thoracic aorta (Fiscus et al. 1991).

Vasoactive intestinal polypeptide is another sensory neuropeptide which, on injection into human skin, causes a local erythema (Anand et al. 1983) that is shorter lasting than that evoked by CGRP (Brain et al. 1986). VIP-induced hyperaemia is also seen in the rabbit skin (T.J. Williams 1982), and arterial infusion of VIP to the rat hindleg enhances venous outflow from this organ (Lembeck and Donnerer 1981a). In contrast, cholecystokinin-like peptides, enkephalins and somatostatin do not affect blood flow in the rat hindleg (Lembeck and Holzer 1979; Lembeck and Donnerer 1981a; Lembeck et al. 1982).

3.2.3.3 Role of CGRP, SP and Other Sensory Neuropeptides in Afferent Nerve-Mediated Vasodilatation

The co-existence of CGRP and SP in perivascular afferent nerve fibres, their release upon nerve stimulation, the presence of SP and CGRP receptors on cutaneous blood vessels, and their highactivity in dilating cutaneous blood vessels form the anatomical and functional substrate for these peptides being the prime candidate mediators of arteriolar dilatation evoked by afferent nerve stimulation. Firm support for an implication of SP and CGRP in cutaneous vasodilatation has come from the use of receptor desensitization, peptide immunoneutralization and peptide antagonists. Antidromic vasodilatation in the rat hindleg is blunted after rendering the preparation insensitive to SP by exposure to a desensitizing dose of the peptide (Lembeck and Holzer 1979). Likewise, tachykinin antagonists have been found to inhibit the vasodilatation induced by SP or antidromic nerve stimulation in the cat's tooth pulp (Rosell et al. 1981), rat hindleg (Lembeck et al. 1981, 1982) and rat cheek (Couture and Cuello 1984).

The finding that tachykinin antagonists fail to completely block the vasodilator response to nerve stimulation indicates that SP and related peptides are not the only mediators. SP is co-released with CGRP from sensory nerve terminals (Yonehara et al. 1987; 1991), and each peptide is likely to contribute to cutaneous vasodilatation. Evidence for a co-mediatorship of SP and CGRP has been provided by immunoblockade of these two peptides. Vasodilatation evoked by topical mustard oil in the rat paw skin is reduced after intravenous injection of CGRP antibodies, but the combined pretreatment wich CGRP and SP antibodies is more effective than CGRP immunoneutralization alone (S.M. Louis et al. 1989a). Similarly, a CGRP antibody Fab fragment (Buckley et al. 1992) and the CGRP antagonist, $CGRP_{8-37}$ (Hughes and Brain 1991), are able to partially antagonize the hyperaemic effect of intradermal capsaicin in the rabbit skin. The vasodilatation induced by topical mustard oil in the rat hindpaw skin is reduced after inhibition of NO synthesis, which suggests that the neural mediators of the vasodilator response act at least in part via release of EDRF (Lippe et al. 1992). There is circumstantial evidence that EDRF also participates in antidromic vasodilatation (Low and Westerman 1989).

Certain sensory neuropeptides such as somastostatin and leucine-enke-phalin/dynorphin, which per se have no effect on blood flow, appear to modulate the vascular effects of sensory nerve stimulation. Thus, somato-statin can inhibit the release of SP from sensory neurons (Brodin et al. 1981; Gazelius et al. 1981) and reduce antidromic vasodilatation in the cat's tooth pulp (Gazelius et al. 1981) and rat hindleg (Lembeck et al. 1982). Opiate agonists are also capable of inhibiting antidromic vasodila-tation in the rat hindleg (Lembeck et al. 1982; Lembeck and Donnerer 1985), rat hindpaw skin (Gamse and Saria 1987) and pig skin (Barthó et al. 1990a), an effect that is possibly due to inhibition of transmitter release from the peripheral endings of afferent nerve fibres (Konishi et al. 1980; Brodin et al. 1983a; Yaksh 1988; Yonehara et al. 1988). It is not known, however, whether afferent nerve fibres containing somatostatin or leucine-enkephalin play any role in the local control of vessel diameter.

3.2.4 Increase in Vascular Permeability

3.2.4.1 Afferent Nerve-Mediated Vasodilatation Versus Protein Extravasation in the Rat Skin

Cutaneous vasodilatation caused by afferent nerve stimulation in the rat is accompanied by an increase in vascular permeability leading to leakage of plasma protein into the interstitial space. Although the basic mechanism of both the exudation and vasodilator response involves the release of me-diator peptides from afferent nerve terminals, there are important differen-ces between the two vascular reactions. Unlike hyperaemia, which occurs by dilatation of arterioles, protein leakage results from an increase in the permeability of postcapillary venules (G. Jancsó 1984; Kenins et al. 1984; Kowalski et al. 1990; Gao et al. 1991) whose endothelial cells contract and allow for formation of gaps in the endothelium (Majno et al. 1969). The spatial separation of the processes underlying vasodilatation and protein extravasation implies that different nerve fibres control the two vascular responses. This contention is supported by differences in the nature of the fibres that give rise to vasodilatation and those responsible for exudation.

3.2.4.2 Nature of Afferent Fibres Mediating Plasma Protein Extravasation

Whereas both C fibres and some Aδ fibres govern arteriolar dilatation, only C fibres give rise to increased venular permeability (Jänig and Lisney 1989). The C fibres which regulate extravasation in the rat paw skin are connected to polymodel nociceptors (Kenins 1981; Bharali and Lisney 1988), yet only part of the polymodal nociceptor fibres cause protein leakage upon stimulation (Bharali and Lisney 1988; Lisney and Bharali 1989). Excitability in relation to stimulus frequency is another property that discriminates between the fibres controlling arteriolar diameter and venular permeability. Unlike vasodilatation, which can be produced by a few shocks (Magerl et al. 1987; Lynn 1988; Lynn and Skakhanbeh 1988a; Szolcsányi 1988), vascular permeability does not rise unless nerve fibres are stimulated at a frequency of 2 Hz or more (Kenins 1981; Szolcsányi 1984; Brenan et al. 1988; Lisney and Bharali 1989; Szolcsányi et al. 1992). Although these differences could in part be accounted for by regional differences in postjunctional mechanisms (e.g. peptidase activity) along the vascular tree, they do point to the possibility that different populations of fine afferent nerve fibres give rise to cutaneous hyperaemia and protein exudation (Lisney and Bharali 1989). Both populations, however, are sensitive to the stimulant and neurotoxic actions of capsaicin. Application of capsaicin to the rat skin leads to a local increase of vascular permeability (Kenins et al. 1984; Carter and Francis 1991), whereas exudative responses to capsaicin in the skin of humans (Helme and McKernan 1984, 1985; Barnes et al. 1986; Lundblad et al. 1987) and pigs (Alving et al. 1991a, b; Pierau et al. 1991) are weak or absent. Functional ablation of capsaicin-sensitive afferent neurons inhibits protein leakage and/or oedema caused by antidromic nerve stimulation (N. Janscó et al. 1967; G. Janscó et al. 1977, 1980; Lembeck and Holzer 1979; Gamse et al. 1980; Morton and Chahl 1980; Lembeck and Donnerer 1981b; Szolcsányi 1988; Andrews et al. 1989; Carter and Francis 1991; Szolcsányi et al. 1992) or local application of mustard oil, KCl, HCl, formalin, antigen, staphylococcal enterotoxin B, prostaglandin E_1, bradykinin, SP, compound 48/80, 5-hydroxytryptamine, histamine and heat (N. Janscó et al. 1967; G. Janscó et al. 1977, 1980, 1985; Arvier et al. 1977; Lembeck and Holzer 1979; Gamse et al. 1980; Lembeck and Donnerer 1981b; Saria et al. 1983, 1984; Saria 1984; Yonehara et al. 1987; Lynn and Shakhanbeh 1988b; Alber et al. 1989) in the skin of rats, rabbits, monkeys and humans.

3.2.4.3 Effects of Sensory Neuropeptides on Vascular Permeability

Among the sensory neuropeptides, tachykinins excel with their unique ability to increase vascular permeability, an activity that was recognized (Starke 1964) before the structure of SP was identified. When injected into the skin of guinea pigs (Woodward et al. 1985; Iwamoto and Nadel 1989), mice (Yano et al. 1989), rats (Chahl 1979; Gamse and Saria 1985b; Devillier et al. 1986a; Maggi et al. 1987a; Yonehara et al. 1987; Brain and Williams 1988, 1989; Devor et al. 1989) or humans (Hägermark et al. 1978; Carpenter and Lynn 1981; Anand et al. 1983; Foreman et al. 1983; Jorizzo et al. 1983; Piotrowski and Foreman 1985; Devillier et al. 1986a; Fuller et al. 1987; Wallengren and Håkanson 1987; Brain and Williams 1988, 1989; Pedersen-Bjergaard et al. 1989, 1991; Iwamoto et al. 1990; Heyer et al. 1991), tachykinins induce plasma protein leakage and local oedema (weal). Extravasation is also seen when tachykinins are superfused over the hamster cheek pouch (Raud et al. 1991) or the base of vacuum-induced blisters in the rat paw skin (Khalil et al. 1988; Andrews et al. 1989; Khalil and Helme 1989b), administered close arterially into the rat hind leg (Lembeck et al. 1977; Lembeck and Holzer 1979; Gamse et al. 1980; Lembeck and Donnerer 1981a; Lembeck et al. 1982; Kenins et al. 1984) or injected subcutaneously (Gao et al. 1991) or intravenously (Saria et al. 1983; Hua et al. 1984; Lundberg et al. 1984a, b; Jacques et al. 1989; Kowalski et al. 1990) to rats or guinea pigs. The SP-induced increase in vascular permeability is confined to postcapillary venules (Kenins et al. 1984; Kowalski et al. 1990; Gao et al. 1991).

At least three different mechanisms are responsible for the exudative action of tachykinins.

1. SP and related peptides can act directly on the blood vessels to augment protein extravasation. This target of action is reflected by the reported inability of atropine, methysergide, bradykinin antagonists (Jacques et al. 1989), histamine antagonists (Saria et al. 1983; Lundberg et al. 1984a; Wallengren and Håkanson 1987; Brain and Williams 1989; Iwamoto and Nadel 1989; Jacques et al. 1989; Khalil and Helme 1989b), indomethacin (Jorizzo et al. 1983) and defunctionalization of capsaicin-sensitive afferent neurons (Gamse et al. 1980; Carpenter and Lynn 1981; Anand et al. 1983; Maggi et al. 1987a) to modify tachykinin-evoked protein leakage.

2. SP can activate mast cells to release histamine and other mediators that further enhance venular permeability. The histamine-dependent component of the exudative response to SP (Hägermark et al. 1978; Chahl 1979; Lembeck and Holzer 1979; Foreman et al. 1983; Jorizzo et al. 1983; Woodward et al. 1985; Fuller et al. 1987; Wallengren and Håkanson 1987; Brain and Williams 1989; Khalil and Helme 1989b) and

NKA (Fuller et al. 1987) is best observed after intradermal injection of the peptide whilst protein leakage evoked by intravascular administration of SP (Saria et al. 1983; Lundberg et al. 1984a; Jacques et al. 1989) does not necessarily involve histamine and other non-neural secondary mediators. Temporal analysis of the exudative effect of SP has shown that histamine, 5-hydroxytryptamine and prostanoids released from mast cells come into play only after some delay, whereas the early phase of exudation is due to a direct vascular action of SP (Chahl 1979; Khalil and Helme 1989b). Consideration of this shift in the extravasation mechanisms provides a possibility to reconcile discrepant observations as to whether or not tachykinin-induced extravasation involves histamine, 5-hydroxytryptamine, prostanoids and capsaicin-sensitive afferent neurons activated by these algesic chemicals. The implication of these secondary mediators in the exudative responses to tachykinins is detailed in section 3.2.6.4.

3. Protein leakage induced by SP involves endothelium-derived NO which, by faciliating local blood flow, may add to the leakage of plasma proteins and fluid (Hughes et al. 1990).

The receptors which mediate the histamine-independent extravasation response to tachykinins seem to be predominantly of the NK_1 type although the implication of NK_2 and NK_3 receptors has not yet been entirely ruled out. Inconclusive evidence for a participation of NK_1 receptors came from the rank order of potency with which different tachykinins and their receptor-selective analogues increase vascular permeability in the skin, SP being either as potent as NKA or more potent than NKA and NKB. This applies both to the weal evoked by tachykinins in human skin (Foreman et al. 1983; Piotrowski et al. 1984; Devillier et al. 1986a; Fuller et al. 1987; Wallengren and Håkanson 1987) and to the tachykinin-induced exudation in rodent skin (Hua et al. 1984; Devillier et al. 1986a; Fuller et al. 1987; Andrews et al. 1989; Devor et al. 1989; Iwamoto and Nadel 1989; Jacques et al. 1989; Khalil and Helme 1989b). However, two studies have found NKB to be more potent than SP in inducing extravasation in the rat skin (Gamse and Saria 1985b; Couture and Kérouac 1987), and cross-tachyphylaxis experiments in the guinea pig skin have been used to implicate NK_1, NK_2 and NK_3 receptors in tachykinin-induced extravasation (Iwamoto and Nadel 1989). The application of receptor-selective tachykinin antagonists has indicated, though, that the exudative effect of SP in the rat skin is mediated by NK_1 receptors (Xu et al. 1992).

Although dynorphin$_{1-13}$ has been reported to induce protein exudation in the rat skin (Chahl and Chahl 1986), other sensory neuropeptides lack consistent activity on cutaneous vascular permeability. Intradermally or intraarterially administered CGRP (Brain and Williams 1985, 1988, 1989;

Gamse and Saria 1985b; Hughes and Brain 1991), and VIP (Chahl 1979; Gamse et al. 1980; Lembeck and Donnerer 1981a; T.J. Williams 1982) do not enhance vascular permeability in the rat or rabbit skin, or are very weakly active in this respect. Topical administration of CGRP to the hamster cheek pouch also fails to induce leakage of plasma proteins (Raud et al. 1991). Cholecystokinin-like peptides, enkephalins, and somatostatin are likewise devoid of actions on vascular permeability in the rat skin (Chahl 1979; Gamse et al. 1980; Barthó and Szolcsányi 1981; Lembeck and Donnerer 1981a; Lembeck et al. 1982). However, a weak weal response to intradermally injected CGRP (Piotrowski and Foreman 1986; Pedersen-Bjergaard et al. 1991), VIP and somatostatin (Anand et al. 1983) has sometimes been noted in human skin.

3.2.4.4 Role of SP in Afferent Nerve-Mediated Protein Extravasation

Although neurochemical studies indicate that SP is not the only mediator (Chahl and Manley 1980; Gamse and Saria 1985a), there is indirect and direct evidence that this tachykinin is a prime mediator of afferent nerve-mediated protein leakage. SP is very potent in increasing vascular permeability in the skin. Postcapillary venules (G. Jancsó 1984; Kenins et al. 1984; Kowalski et al. 1990; Gao et al. 1991) which bear NK_1 receptors (Deguchi et al. 1989; O'Flynn et al. 1989) are the common target of action of antidromic nerve stimulation and SP in increasing vascular permeability. The exudative responses to antidromic nerve stimulation and SP are similar with regard to pharmacology (Chahl 1979; Lembeck and Holzer 1979; Morton and Chahl 1980; Khalil and Helme 1989b) and time course inasmuch as the early phase of extravasation is due to a direct action on venular endothelial cells whilst later phases involve secondary mediators such as histamine, 5-hydroxytryptamine and prostanoids (Chahl 1979; Morton and Chahl 1980; Kowalski and Kaliner 1988; Khalil and Helme 1989b; Kowalski et al. 1990).

Direct evidence for a mediator role of SP comes from studies using pharmacological antagonism of the peptide's action. Tachykinin antagonists inhibit cutaneous plasma protein exudation evoked by SP (Lembeck et al. 1982, 1992; Chahl et al. 1984; Lundberg et al. 1984b; Khalil and Helme 1989b; Xu et al. 1991), antidromic nerve stimulation (Lembeck et al. 1982; Chahl et al. 1984; Couture and Cuello 1984; Lundberg et al. 1984b; Xu et al. 1991), noxious heat (Lundberg et al. 1984b; Saria 1984) and cutaneous application of capsaicin (Mantione and Rodriguez 1990), bradykinin (Shibata et al. 1986; Mantione and Rodriguez 1990) or staphylococcal enterotoxin B (Alber et al. 1989). The use of receptor-selective antagonists has shown hat the afferent nerve-mediated increase in vascular permeability is mediated by the NK_1 tachykinin receptor type (Eglezos et al. 1991; Garret et al. 1991; Lembeck et al. 1992; Xu et al. 1992). The inhibitory effect of tachykinin antagonists is matched by the ability of SP antibodies to reduce the exudative responses to antidromic nerve stimulation, topical mustard oil (S.M. Louis et al. 1989a) and intradermal staphylococcal enterotoxin B (Alber et al. 1989). Sensory nerve-mediated increases in vascular permeability in the eye (Holmdahl et al. 1981), joints (Ferrell and Russell 1986) and respiratory tract (see Lundberg and Saria 1987; Barnes et al. 1988; Lembeck et al. 1992) are also attenuated by SP antagonists. In the rat cheek skin it appears as if SP antagonists are more active in reducing the extravasation than the vasodilatation caused by antidromic nerve stimulation (Couture and Cuello 1984).

3.2.4.5 Role of CGRP and Other Sensory Neuropeptides in Afferent Nerve-Mediated Protein Extravasation

The precise role of afferent nerve-derived CGRP in the control of vascular permeability is still elusive because this peptide is able to exert both proinflammatory and antiinflammatory effects. The proinflammatory action of CGRP is reflected by the peptide's ability to enhance protein leakage in response to tachykinins (Brain and Williams 1985, 1988, 1989; Gamse and Saria 1985b), bradykinin, platelet-activating factor, N-formyl-methionyl-leucyl-phenylalanine and a factor of the complement system (Buckley et al. 1991) in the rat and rabbit skin. This CGRP-induced facilitation of protein extravasation is thought to result from its vasodilator activity (Buckley et al. 1991; Hughes and Brain 1991), although inhibition of SP degradation by CGRP (Le Grevès et al. 1985, 1989) may also play a role. These interactions between CGRP and SP may also explain why CGRP seems to be a "mediator" of afferent nerve-mediated protein extravasation in the rat skin, although this peptide per se is without effect on vascular permeability. Intravenous injection of CGRP antibodies, like that of SP antibodies, reduces extravasation in the rat skin caused by antidromic stimulation of the saphenous nerve or cutaneous application of mustard oil (S.M. Louis et al. 1989a). The response to topical mustard oil is also diminished by active immunization of rats against CGRP (S.M. Louis et al. 1989b). An implication of both SP and CGRP in neurogenic plasma protein leakage is furthermore consistent with the finding that, following nerve injury and regeneration, the ability of capsaicin or mustard oil to enhance vascular permeability is highly correlated with the levels of both SP and CGRP in the skin area under study (McMahon et al. 1989; see also Lundberg et al. 1984a).

Calcitonin gene-related peptide-induced potentiation of the exudative response to SP is not seen in the human skin (Wallengren and Håkanson 1984; Pedersen-Bjergaard et al. 1991), and exudative responses to histamine, 5-hydroxytryptamine and leukotriene B_4 in the hamster cheek pouch, rat hindpaw and human forearm skin are in fact reduced by CGRP (Raud et al. 1991). This antiinflammatory effect of CGRP is mimicked by acute administration of capsaicin (Raud et al. 1991). It is worth nothing in this respect that antidromic stimulation of dorsal roots releases an unidentified factor that has systemic antiexudative activity in the rat (Szolcsányi and Pintér 1991).

Apart from CGRP, there are other sensory neuropeptides which per se do not evoke protein leakage but modulate the exudative responses to anti-dromic nerve stimulation or SP administration. VIP augments protein leakage in response to SP, bradykinin and a factor of the complement system in the rat and rabbit skin (T.J. Williams 1982; Khalil et al. 1988), an effect that is probably related to the vasodilator activity of VIP. Enke-phalins (Morton and Chahl 1980; Barthó and Szolcsányi 1981; Lembeck et al. 1982; Smith and Buchan 1984; Yonehara et al. 1988) are able to di-minish afferent nerve-mediated protein exudation, possibly through inhi-bition of neuropeptide release from afferent nerve terminals (Konishi et al. 1980; Brodin et al. 1983a; Yaksh 1988; Yonehara et al. 1988). A similar mechanism of action (Brodin et al. 1981; Gazelius et al. 1981) may ac-count for the inhibitory effect of somatostatin on afferent nerve-mediated extravasation in the skin (Lembeck et al. 1982). Corticotropin-releasing factor (Kiang and Wei 1985; Gao et al. 1991) and galanin (Xu et al. 1991) are able to reduce protein leakage evoked by both antidromic nerve stimu-lation and SP, which indicates a postjunctional target of action. Taken to-gether, these data suggest that the SP-mediated exudative response to affe-rent nerve stimulation is modulated by other sensory neuropeptides which are co-released with SP.

3.2.5 Interaction of Afferent Nerve Fibres with Autonomic Neurons in the Control of Blood Flow and Vascular Permeability

3.2.5.1 Cholinergic Neurons

Acytylcholine is unlikely to be a primary mediator of afferent nerve-mediated vasodilatation and protein exudation because atropine fails to block, and eserine fails to potentiate, antidromic vasodilatation in the rab-bit ear (Holton and Perry 1951). The hyperaemic response of rat skin to antidromic or intradermal nerve stimulation either remains unchanged (Gamse and Saria 1987) or is reduced (Lembeck and Holzer 1979; Couture and Cuello 1984; Couture et al. 1985; Low and Westerman 1989) by atro-pine. Similarly, afferent nerve-mediated protein leakage is left unaltered (N. Jancsó et al. 1967) or is attenuated (Couture and Cuello 1984; Couture et al. 1985) by atropine. The inhibitory effect of atropine has been explained in two different ways: (a) stimulation of mixed nerves is likely to activate efferent cholinergic nerve fibres (i.e. postganglionic parasympathetic fibres, sympathetic sudomotor fibres), acetylcholine contributing to the vascular effects of nerve stimulation (Couture and Cuello 1984; Couture et al. 1985; Low and Westerman 1989); (b) acetylcholine may play some role as a secondary mediator of SP (Tanaka and Grunstein 1985), as atropine reduces the exudative response to intradermal SP in the rat dorsal skin (Couture and Kérouac 1987), whereas the exudative response to intravenous SP is not altered (Jacques et al. 1989).

3.2.5.2 Noradrenergic Sympathetic Neurons

The available information indicates that noradrenergic sympathetic neurons do not contribute to neurogenic vasodilatation and protein leakage in the skin. Antidromic stimulation of ventral roots (L4-S1) fails to evoke hyperaemia and protein exudation in the rat hindpaw skin, whilst antidromic stimulation of the saphenous nerve induces vasoconstriction if afferent neurons have been defunctionalized by capsaicin (Szolcsányi et al. 1992). This vasoconstrictor response is due to stimulation of noradrenergic fibres since it is blocked by phentolamine and guanethidine (Szolcsányi et al. 1992). In addition, protein extravasation in the paw skin evoked by antidromic stimulation of the saphenous nerve or topical application of mustard oil remains unaltered after a 60%–86% depletion of noradrenaline from the skin as induced by pretreatment of rats with guanethidine or 6-hydroxydopamine (Donnerer et al. 1991). In contrast, in the joint a cascade of events in which the sensory neuropeptide substance P liberates mast cell histamine which in turn releases prostaglandins from sympathetic nerve endings and thereby amplifies protein extravasation has been proposed (Basbaum and Levine 1991). This concept is based, among others, on the finding that articular protein leakage induced by local administration of capsaicin (5 mg ml^{-1}) is reduced in rats pretreated with 6-hydroxydopamine (Coderre et al. 1989). However, the exudative response to intraarticular capsaicin as seen in the study of Coderre et al. (1989) was only in part due to stimulation of afferent neurons because a considerable proportion of the response persisted after ablation of capsaicin-sensitive afferent neurons. In thus remains unclear whether 6-hydroxydopamine blocked the afferent nerve-dependent or -independent component of capsaicin-evoked protein extravasation. In addition, sympathetic denervation could lead to changes in the microcirculation and thereby interfere with vascular inflammatory reactions (Lembeck et al. 1977; Osswald 1990; Donnerer et al. 1991).

There is consistent evidence, though, that sympathetic noradrenergic neurons counteract afferent nerve-mediated vasodilatation (Hornyak et al.

1990) and protein extravasation (Fearn et al. 1965; Helme and Andrews 1985; Lindgren et al. 1987) in the skin. Hyperaemia in the rat hindleg due to antidromic stimulation of the saphenous nerve is in fact only seen when noradrenergic nerve fibres have been blocked by guanethidine (Lembeck and Holzer 1979), and antidromic vasodilatation in the rat hindpaw skin is amplified by guanethidine pretreatment (Gamse and Saria 1987). The finding that activation of α_2 adrenoceptors (Lindgren et al. 1987) also inhibits afferent nerve-mediated hyperaemia and protein leakage in the skin may point to a prejunctional inhibition of transmitter release from afferent nerve terminals as has been seen in other tissues (see Maggi 1991). In the central artery of the rabbit ear, the interaction between afferent and noradrenergic sympathetic neurons seems to be reciprocal because CGRP can inhibit noradrenergic transmission and thereby lessen the sympathetic vasoconstrictor tone (Maynard et al. 1990).

3.2.6 Non-Neural and Non-Vascular Secondary Mediators of Afferent Nerve-Mediated Vasodilatation and Protein Extravasation

3.2.6.1 Association of Afferent Nerve Fibres with Mast Cells

There are histochemically demonstrable contact sites between sensory nerve endings and mast cells in the human skin (Naukkarinen et al. 1991). Some nerve fibres immunoreactive for CGRP and SP in the pig's skin are in close association with histamine-containing mast cells around arteries, arterioles and venules (Alving et al. 1991b). Such a close apposition of peptidergic afferent nerve endings and mast cells has also been observed in non-cutaneous tissues (Skofitsch et al. 1985a; Stead et al. 1987b, 1989).

3.2.6.2 Involvement of Mast Cell Mediators in the Postacute Phase of the Vascular Effects of Afferent Nerve Stimulation

The participation of histamine and other mediators released from mast cells in afferent nerve-mediated vasodilatation and protein extravasation has long been disputed. Part of the controversy has arisen from failure to consider that mast cell-derived histamine is involved in the protacted stages of neurogenic inflammation, whilst the initial stages result from a direct action of sensory neuropeptides on cutaneous blood vessels (Morton and Chahl 1980; Kowalski and Kaliner 1988; Kowalski et al. 1990). When examined 3 min after antidromic stimulation of the rat saphenous nerve, protein leakage takes place primarily in the superficial dermis (Kowalski et al. 1990) where peptidergic afferent nerve fibres abound (Hökfelt et al. 1975; Cuello et al. 1978; Dalsgaard et al. 1983; Hartschuh et al. 1983; Gibbins et al. 1985, 1987b; Wallengren et al. 1987). In contrast, most skin mast cells are located in deeper layers of the dermis (Kowalski et al. 1990), and no sign of mast cell degranulation or histamine depletion from the skin becomes discernible during the early phase of plasma leakage due to anti-dromic nerve stimulation (Kowalski and Kaliner 1988). The finding is consistent with the failure of the mast cell stabilizer disodium cromogly-cate (Kowalski et al. 1990) and of histamine H_1 antagonists to block the onset of the hyperaemic (Kiernan 1976; Gamse and Saria 1987) and exudative (N. Jancsó et al. 1967; Garcia Leme and Hamamura 1974; Morton and Chahl 1980; Lundberg et al. 1984a, b) response to antidromic nerve stimulation. Prostanoids and 5-hydroxytryptamine, two further mast cell mediators, are also unlikely to play a role in the initial stages of neuro-genic inflammation because indomethacin (Lembeck and Holzer 1979; Morton and Chahl 1980; Couture and Cuello 1984) and methysergide (Lembeck and Holzer 1979) do not change vasodilatation and extravasa-tion induced by antidromic nerve stimulation. As might be expected from these observations, the first phase of protein extravasation caused by anti-dromic nerve stimulation is unchanged in mice that are genetically mast cell deficient (Kowalski et al. 1990).

When antidromic stimulation of the saphenous nerve is continued for 30 min, mast cells in the rat paw degranulate and the cutaneous histamine content is significantly lowered (Kowalski and Kaliner 1988). Mast cell degranulation also occurs in the rat ear following antidromic stimulation of the great auricular nerve (Kiernan 1971) and in the region of the arteriolar flare surrounding a cutaneous scratch (Kiernan 1972, 1984). Conversely, compound 48/80-induced degranulation of mast cells reduces afferent nerve-mediated vasodilatation (Kiernan 1975; Lembeck and Holzer 1979) and exudation (Arvier et al. 1977; Lembeck and Holzer 1979; Coderre et al. 1989). These observations fit with pharmacological evidence that histamine can contribute to the vasodilator (Lembeck and Holzer 1979; Couture and Cuello 1984; Low and Westerman 1989) and exudative (Lembeck and Holzer 1979; Morton and Chahl 1980; Couture and Cuello 1984) responses to afferent nerve stimulation. Mast cell-derived 5-hydro-xytryptamine (Morton and Chahl 1980; Couture and Cuello 1984) and prostaglandins (Coderre et al. 1989) are also implicated in the protraction of neurogenically initiated inflammation.

3.2.6.3 Effects of Sensory Neuropeptides on the Release of Histamine and Other Factors from Mast Cells: Tissue and Species Differences

Congruous with the involvement of mast cell-derived factors in neurogenic inflammation is the ability of sensory neuropeptides to release histamine from mast cells, an activity that shows marked tissue and species differences. SP is able to release histamine from peritoneal mast cells of rat

(Johnson and Erdös 1973; Erjavec et al. 1981; Fewtrell et al. 1982; Håkanson et al. 1983; Piotrowski et al. 1984; Shanahan et al. 1985; Ali et al. 1986; Devillier et al. 1986a; Pearce et al. 1989), mouse and hamster (Ali et al. 1986; Pearce et al. 1989), whereas human peritoneal mast cells do not respond (Ali et al. 1986). However, human skin mast cells release histamine in response to SP, VIP and somatostatin (Ebertz et al. 1987; Lowman et al. 1988; Benyon et al. 1989; Church et al. 1989), whilst mast cells from other human tissues so far studied fail to react (Ali et al. 1986; Church et al. 1989; Pearce et al. 1989). Also VIP and somatostatin are active on rat peritoneal mast cells (Theoharides and Douglas 1978; Shanahan et al. 1985; Pearce et al. 1989) which therefore have a similar spectrum of neuropeptide sensitivity as human skin mast cells (Lowman et al. 1988; Church et al. 1989). This similarity is further underlined by the inactivity of NKA and NKB to induce histamine release from rat peritoneal (Piotrowski et al. 1984; Devillier et al. 1986a) and human skin (Lowman et al. 1988) mast cells. In contrast, CGRP can activate rat peritoneal mast cells (Piotrowski and Foreman 1986) but is rather inactive on human skin mast cells (Lowman et al. 1988; Church et al. 1989). Mast cells from a variety of guinea pig tissues do not respond to SP (Ali et al. 1986; Pearce et al. 1989), and mucosal mast cells from the rat intestine respond to SP but not somatostatin or VIP (Shanahan et al. 1985; Pearce et al. 1989). In line with its activity on isolated human skin mast cells, SP also releases histamine in the human skin in vivo (Barnes et al. 1986) and in vitro (Ebertz et al. 1987; R.E. Louis and Radermecker 1990). Similarly, the release of histamine from the isolated rat hindleg is augmented by SP (Skofitsch et al. 1983; Holzer-Petsche et al. 1985), whereas NKA (Holzer-Petsche et al. 1985) is inactive. Somatostatin and VIP, but not SP, also release 5-hydroxytryptamine in this preparation (Skofitsch et al. 1983).

Unlike IgE, SP is only weakly active in releasing prostaglandin D_2 and leukotriene C_4 from human skin mast cells (Benyon et al. 1989).

3.2.6.4 Involvement of Mast Cell Mediators in the Vascular Effects of Sensory Neuropeptides

The ability of SP to release histamine from skin mast cells has a bearing on the peptide's vascular effects which may in part be mediated by mast cell-derived factors. However, there are tissue and species differences in the relative importance of mast cells in tachykinin-evoked inflammation. As with the vascular effects of afferent nerve stimulation, mast cell mediators such as histamine play an important role in the later phases of SP-induced hyperaemia and protein leakage only. Several lines of evidence support this contention.

1. The first phase of protein extravasation caused by intravenous administration of SP remains unaltered in mice which are genetically mast cell deficient (Kowalski et al. 1990). Two hours after SP injection, however, the increase in vascular permeability is greatly reduced in these animals (Yano et al. 1989).

2. These findings are consistent with pharmacological studies in which antagonists of mast cell mediators were found to inhibit the delayed phase of SP's vascular effects whereas the early phase was left unaltered (Chahl 1979; Khalil and Helme 1989b). Lack of consideration of this temporal shift in the mast cell-independent and -dependent mechanisms of SP's actions on blood vessels may be an important reason for the controversy over the role of histamine as a secondary mediator of SP.

3. Degranulation of mast cells by pretreatment of rats with compound 48/80 has been found to inhibit SP-induced hyperaemia and protein extravasation in the rat hindleg (Lembeck and Holzer 1979).

4. Both the vasodilator (Lembeck and Holzer 1979; Andrews and Helme 1989) and exudative (Hägermark et al. 1978; Chahl 1979; Lembeck and Holzer 1979; Foreman et al. 1983; Jorizzo et al. 1983; Woodward et al. 1985; Fuller et al. 1987; Wallengren and Håkanson 1987; Brain and Williams 1989; Khalil and Helme 1989b) effects of SP can in part be inhibited by histamine antagonists, sometimes in combination with methysergide (Chahl 1979). In the guinea pig skin, histamine appears to play a more important role in the exudative effects of SP than in the vasodilator effects of the peptide (Woodword et al. 1985).

5. In addition to histamine and 5-hydroxytryptamine, mast cell-derived prostanoids also seem to mediate a component of the vascular actions of SP, but not NKA and NKB (Couture et al. 1989; Jacques et al. 1989; Khalil and Helme 1989b).

6. Similarly, SP leads to release of proteases from mast cells in rat and human skin whilst NKA and NKB are inactive. The SP-released proteases are able to cut short the prolonged vasodilator activity of CGRP (Brain and Williams 1988, 1989), which indicates that mast cell constituents play an important role in the interactive regulation of the vascular effects of co-released sensory neuropeptides.

7. The findings that the vascular effects of SP in the skin are reduced by local anaesthetics (Foreman et al. 1983; Wallengren and Håkanson 1987) and defunctionalization of capsaicin-sensitive afferent neurons (Chahl and Chahl 1986; G. Jancsó et al. 1985; Andrews and Helme 1989; Devor et al. 1989) may be explained by SP-induced release of mast cell-derived histamine and 5-hydroxytryptamine which in turn can activate afferent neurons (see Lang et al. 1990) and thus give rise to afferent nerve-mediated vasodilatation (Khalil and Helme 1989a) and

protein exudation (see G. Jancsó et al. 1980, 1985; Saria et al. 1983, 1984; G. Jancsó 1984). The observation that afferent neurons are not required for protein leakage caused by Arg-NKB (Maggi et al. 1987a), which is unlikely to release histamine from mast cells (Piotrowski et al. 1984; Devillier et al. 1986a; Lowman et al. 1988), supports this argument.

3.2.6.5 Nature of the Mast Cell Receptors for SP

There is an important difference between the SP receptors which mediate the peptide's mast cell-independent and -dependent effects on blood flow and vascular permeability. The histamine-independent effects of SP and NKA to induce vasodilatation (Holzer-Petsche et al. 1985; Maggi et al. 1985; Andrews and Helme 1989; Couture et al. 1989) and plasma exudation (Devillier et al. 1986a; Fuller et al. 1987; Andrews et al. 1989; Devor et al. 1989; Iwamoto and Nadel 1989; Jacques et al. 1989; Khalil and Helme 1989b) are predominantly mediated by NK_1 receptors which recognize certain features of the C terminal amino acid sequence of tachykinins. In contrast, the mast cell-dependent effects of SP are determined by certain structural features of the N-terminal part of the molecule in which the presence of basic amino acids plays a critical role (Mazurek et al. 1981; Fewtrell et al. 1982; Foreman et al. 1983; Skofitsch et al. 1983; Piotrowski et al. 1984; Lowman et al. 1988; Church et al. 1989; Khalil and Helme 1989b; Pearce et al. 1989). The structural requirements of a tachykinin to be recognized by the "mast cell receptor" for SP are thus fundamentally disparate from those needed for binding to the typical tachykinin (NK_1, NK_2 and NK_3) receptors. This divergence explains why NKA and NKB are inactive on mast cells (Holzer-Petsche et al. 1985; Devillier et al. 1986a; Lowman et al. 1988; Church et al. 1989; Pearce et al. 1989) and do not mimic the mast cell-dependent actions of SP on blood vessels (Devillier et al. 1986a; Brain and Williams 1988, 1989). The finding that the vascular effects of SP, but not NKA and NKB, can be reduced by indomethacin indicates that SP-induced release of prostanoids is also determined by the N terminus of the peptide (Jacques et al. 1989).

The different nature of the mast cell receptor for SP is underlined by the absence of autoradiographically demonstrable NK_1 receptor sites from mast cells in the rat footpad skin (O'Flynn et al. 1989). Biochemical evidence suggests that SP's action on mast cells is not mediated by a proper membrane receptor for the peptide but results from a direct interaction of SP with G proteins in the mast cell membrane (Bueb et al. 1990). This structural disparity manifests itself in a greatly different affinity for the ligand. Whereas the NK_1 receptor has a dissociation constant in the nanomoles per litre range it is micromoles per litre concentrations of SP which are needed to activate mast cells. It is therefore conceivable that mast cell histamine is not a major determinant in the vascular effects of endogenously released tachykinins but comes into play when relatively high doses of exogenous SP are administered. This argument finds support in the observation that the exudative response to intravenous SP takes place primarily deep in the dermis where mast cells abound while plasma leakage to antidromic nerve stimulation prevails in the superficial dermis (Kowalski et al. 1990).

3.2.7 Afferent Nerve Fibres and the Flare Response

3.2.7.1 The Axon Reflex Hypothesis

Focal irritation of human skin causes (a) a local reddening; (b) an area of oedema (the weal) at the site of the stimulus; and (c) a spread of arteriolar dilatation (the flare) far beyond the point of irritation. The flare component of this "triple response" (Lewis 1927) requires an intact sensory innervation and is abolished by defunctionalization of capsaicin-sensitive afferent neurons in human (N. Jancsó et al. 1968; Bernstein et al. 1981; Carpenter and Lynn 1981; Anand et al. 1983; Foreman et al. 1983; Tóth-Kása et al. 1986; Lundblad et al. 1987; Szolcsányi 1988; Bjerring and Arendt-Nielsen 1990; Simone and Ochoa 1991) and porcine (Pierau and Szolcsányi 1989; Barthó et al. 1990a) skin. Since the spread of flare involves nerve conduction, it is commonly thought to be the result of an "axon reflex" (Chahl 1988; Holzer 1988; Lynn 1988; Lisney and Bharali 1989). This term denotes a reflex that takes place entirely within the arborizations of a single nerve axon (Lewis 1927). When one axon branch is activated by an irritant stimulus, nerve impulses will travel not only centrally, but at the branching point will also pass antidromically in the other branches which may happen to come close to some arterioles (Fig. 3). If so, the periarteriolar branches may release vasodilator transmitters and thereby cause arteriolar dilatation.

The axon reflex hypothesis offers an elegant way of explaining the spreading flare – yet its neurophysiological mechanism is not understood. Although afferent nerve fibres arborize in the skin, with nerve endings both in the epidermis and along blood vessels, it is not clear whether the epidermal and perivascular branches are collaterals of the same afferent nerve axons. The axon reflex concept implies that the area of flare is determined by the size of "neurovascular units" made up by the collateral networks (receptive fields) of individual afferent nerve fibres and the area of arterioles innervated by these collateral networks (Lewis 1927; Helme and McKernan 1984, 1985). However, the area of flare differs greatly in different body regions, ranging from 0.6 to 15 cm^2 (Helme and McKernan

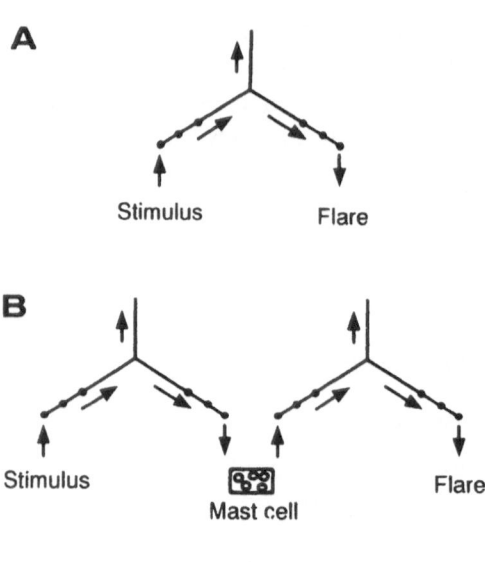

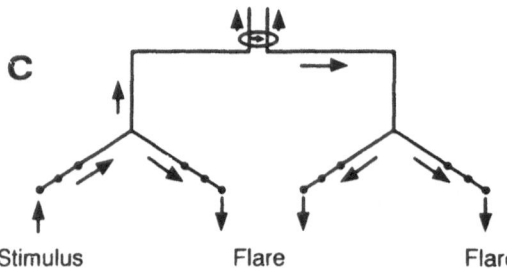

Fig. 3 A–C. Three different axon reflex concepts. **A** Classical axon reflex concept in which an irritative or noxious stimulus activates a collateral of an afferent nerve fibre and nerve activity spreads to all other collaterals. Release of vasodilator mediators at their endings evokes arteriolar dilatation (flare). **B** Chemical coupling between collaterals of two different afferent nerve fibres. Chemical coupling is achieved by release of mast cell-derived histamine and other factors that are able to activate adjacent sensory nerve endings and thus allow nerve activity and flare to spreed beyond the collateral network of a single afferent nerve fibre. **C** Electrical coupling between two different afferent nerve fibres which represents another way by which the flare response may spread beyond the collateral network of a single afferent nerve fibre

1984, 1985), whereas the size of C fibre-receptive fields in human skin is hardly larger than 1 cm^2 (e.g. Torebjörk 1974; van Hees and Gybels 1981; Nordin 1990). This mismatch between flare sizes and C fibre-receptive fields (Lynn 1988) is paralleled by an unexplained delay in the spread of flare which is slower than would be expected if the rate of spread were determined only by conduction delays in unmyelinated nerve fibres (Lynn and Cotsell 1991). To resolve these discrepancies, chemical (see Lembeck and Gamse 1982; Lynn 1988; Lynn and Cotsell 1991) or electrical coupling between different afferent nerve fibres (see B. Matthews 1976; Meyer et al. 1985) or between afferent and efferent sympathetic axons (see Zimmermann 1979) has been considered to participate in the spread of arteriolar flare (Fig. 3). However, coupling must also be limited because the flare has a relatively sharp margin and does not cross the midline of the forehead (Helme and McKernan 1984, 1985).

3.2.7.2 Flare Induced by Sensory Neuropeptides

Like histamine, intradermally injected SP leads to arteriolar flare which is inhibited by histamine H_1 antagonists (Hägermark et al. 1978; Coutts et al. 1981; Jorizzo et al. 1983; Piotrowski and Foreman 1985; Barnes et al. 1986; Fuller et al. 1987; Wallengren and Håkanson 1987; Brain and Williams 1988, 1989; Pedersen-Bjergaard et al. 1989, 1991; Iwamoto et al. 1990; Heyer et al. 1991). It might be concluded from this finding that histamine is a vasodilator mediator of the axon reflex, but there are several lines of evidence indicating that histamine is primarily involved in the initiation of the axon reflex.

1. Unlike SP, NKA and NKB are only weakly active in causing flare (Foreman et al. 1983; Devillier et al. 1986a; Fuller et al. 1987; Wallengren and Håkanson 1987; Pedersen-Bjergaard et al. 1989; Iwamoto et al. 1990), and the relative potencies of tachykinins in evoking flare are similar to those in releasing histamine from human skin (Lowman et al. 1988) and rat peritoneal (Fewtrell et al. 1982; Piotrowski et al. 1984; Devillier et al. 1986a) mast cells. This indicates that the SP-induced flare response is mediated by a "mast cell type" of SP receptor, binding to which depends on the N-terminal amino acid sequence of the ligand.
2. Also VIP and somatostatin liberate histamine from human skin mast cells (Lowman et al. 1988) and elicit flare (Anand et al. 1983). In contrast, the potency of CGRP in inducing flare (Piotrowski and Foreman 1986; Fuller et al. 1987; Wallengren and Håkanson 1987; Pedersen-Bjergaard et al. 1991) and releasing histamine from mast cells (Lowman et al. 1988) is low. The flare responses to VIP, somatostatin (Anand et al. 1983) and CGRP (Piotrowski and Foreman 1986; Fuller et al. 1987; Pedersen-Bjergaard et al. 1991) are inhibited by histamine H_1 antagonists; that to CGRP is also reduced by acetylsalicylic acid (Fuller et al. 1987).
3. Axon reflex flare initiated by electrical stimulation (Parrot 1942) or topical capsaicin administration (Barnes et al. 1986) is not inhibited by histamine antagonists. Since capsaicin does not release histamine (Foreman et al. 1983; Skofitsch et al. 1983), it can be inferred that

histamine does not participate in the production of arteriolar dilatation
in response to an axon reflex (Parrot 1942; Barnes et al. 1986). SP,
though, may be a vasodilator mediator because intradermal injection of
a SP antagonist is able to antagonize SP-induced flare (Foreman and
Jordan 1984). Taken together, the flare response to intradermally injected SP and other sensory neuropeptides is initiated by histamine release
from mast cells, histamine in turn activating sensory nerve terminals
which give rise to the spread of flare.

3.2.8 Involvement of Afferent Neurons in Reactive Hyperaemia

The question whether capsaicin-sensitive afferent neurons contribute to the
vasodilatation following a period of arterial occlusion has not yet been
settled. Defunctionalization of capsaicin-sensitive afferent neurons has
been found to reduce postocclusive hyperaemia in the rat hindleg (Lembeck and Donner 1981a) and gut (Hottenstein et al. 1992), whereas in the
pig postocclusive vasodilatation in skin, skeletal muscle and heart is independent of capsaicin-sensitive afferent neurons (Franco-Cereceda and
Lundberg 1989). In contrast, the cutaneous hyperaemia in the human forearm skin evoked by exercise is blunted 24 h after intradermal injection of
capsaicin (Kurozawa et al. 1991).

3.2.9 Control of Leucocyte Activity by Afferent Nerve Fibres: Significance for Blood Flow and Vascular Permeability

Vasodilatation and increase in vascular permeability are readily visible
consequences of afferent nerve stimulation. Neuropeptides released from
afferent nerve terminals also govern the activity of granulocytes, monocytes and lymphocytes, but these effects are considered here only with regard
to their possible relevance for the vascular aspect of neurogenic inflammation (Fig. 4). Intravascularly, SP induces aggregation of leucocytes and
platelets in skeletal muscle arterioles of the rabbit (Öhlén et al. 1988,
1989), an effect that is antagonized by CGRP (Öhlén et al. 1988). Further
on, SP exerts a chemotactic action on monocytes (Ruff et al. 1985) and
granulocytes (Helme et al. 1987), and stimulates the adhesion of leucocytes to the vessel wall and their emigration into the inflamed tissue of the
skin (Helme and Andrews 1985; Yano et al. 1989; Umeno et al. 1990).
CGRP also causes granulocytes to infiltrate human skin (Piotrowski and
Foreman 1986), whilst in rabbit skin CGRP is without effect on its own
but potentiates granulocyte accumulation initiated by other stimuli
(Buckley et al. 1991). Most important in the present context is SP's ability
to release cytokines (Lotz et al. 1988), prostaglandins, leukotrienes and
thromboxanes (Hartung et al. 1986) from monocytes and histamine (Ali

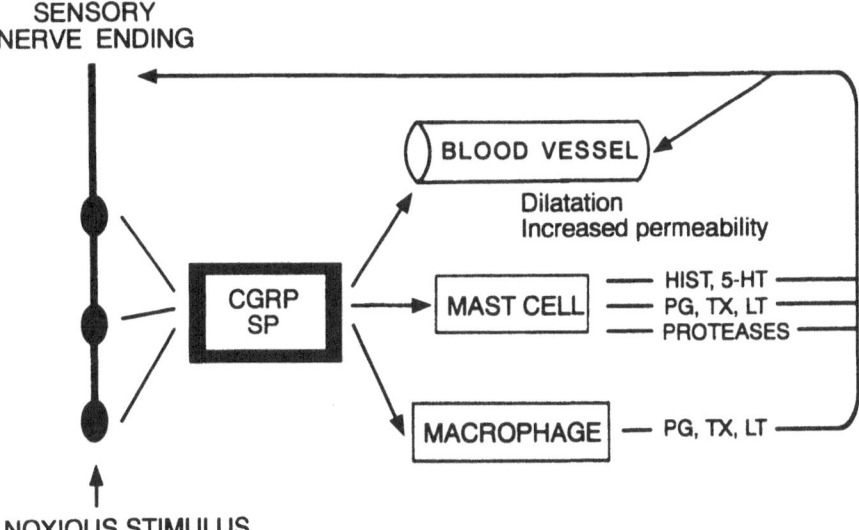

Fig. 4. Reinforcement of the neurogenic inflammatory response to afferent nerve stimulation by mediators released from mast cells and leucocytes, particularly monocytes/macrophages. Many of these mediators are not only vasoactive per se but are capable of sensitizing or stimulating sensory nerve endings. Mast cell-derived proteases can modify the vascular actions of the peptides released from afferent nerve endings (Brain and Williams 1988, 1989). *HIST*, histamine; *5-HT*, 5-hydroxytryptamine; *LT*, leukotrienes; *PG*, prostaglandins; *TX*, thromboxanes

et al. 1986; R.E. Louis and Radermecker 1990) from basophil granulocytes. These leucocyte-derived factors are not only able to influence vessel diameter and permeability but can also stimulate afferent nerve fibres or augment their excitability (see Martin et al. 1987; Maggi 1991) and thus reinforce afferent nerve-mediated hyperaemia and protein leakage (Fig. 4).

Other leucocyte-derived mediators suppress afferent nerve activity and neurogenic inflammation, as do opioid peptides released from immunocytes infiltrating the inflamed tissue (Stein et al. 1990).

3.3 Autonomic Reflex Regulation of Cutaneous Blood Flow

In addition to their role in the local regulation of cutaneous microcirculation, afferent neurons also participate in the reflex regulation of the cardiovascular system. Cutaneous vasodilatation is an important mechanism of thermoregulation designed to facilitate the dissipation of heat. Capsaicin-sensitive afferent neurons comprise warmth receptors (see Szolcsányi 1982, 1983, 1990a), and stimulation of these receptors by capsaicin induces cutaneous hyperaemia whilst body core temperature is lowered (Dib

1983; Donnerer and Lembeck 1983; Hajós et al. 1983; Szolcsányi 1983). These thermoregulatory reflex reactions depend on the sympathetic nervous system inasmuch as cutaneous vasodilatation appears to arise from reflex withdrawal of sympathetic vasoconstrictor tone (Donnerer and Lembeck 1983).

The organism's reactions to nociception are also associated with distinct changes in cardiac performance and regional blood flow. Local intravascular injection of the irritant capsaicin to rats causes a fall in blood pressure (Donnerer and Lembeck 1982). A similar hypotensive response is seen in the rabbit when algesic chemicals such as bradykinin are injected into the ear artery (Juan and Lembeck 1974) which is innervated by capsaicin-sensitive peptidergic afferent neurons (Maynard et al. 1990). In contrast, hypertension and tachycardia ensue after administration of capsaicin to the epicardium (Staszewska-Woolley et al. 1986) or into the skeletal muscle (Crayton et al. 1981) of the dog. The extent to which cutaneous circulation is involved in these cardiovascular reactions to pain is not known, but it is textbook knowledge that nociception is accompanied by sympathetically mediated vasoconstriction in the skin. The same uncertainty applies to the reported participation of capsaicin-sensitive afferent neurons in baroreceptor and chemoreceptor reflex regulation of the circulatory system (Bond et al. 1982; Donnerer et al. 1989).

3.4 Pathophysiological Implications

Dysfunction of peptidergic afferent neurons in the regulation of cutaneous blood flow and vascular permeability is likely to have a bearing on the pathophysiology of the skin. Two pathological changes of these neurons are briefly considered here: malfunction and hyperactivity. Afferent nerve-mediated vasodilatation and axon reflex flare in the skin can be impaired in patients suffering from congenital sensory neuropathy (Tóth-Kása et al. 1984) or sensory neuropathies caused by diabetes (Aronin et al. 1987; Benarroch and Low 1991; Walmsley and Wiles 1991), herpes zoster (Lewis and Marvin 1927; G. Jancsó et al. 1983; Tóth-Kása et al. 1984), postherpetic neuralgia (LeVasseur et al. 1990) or atopic dermatitis (Heyer et al. 1991). Healing of injured tissue requires adequate organizational and functional reactions of the microcirculatory system, and insufficient or improper coordination of these processes may be expected to predispose to "trophic" disorders of the skin. Support for this conjecture comes from clinical observations that sensory neuropathies may be associated with persistent skin ulcers (Böckers et al. 1989) and connective tissue diseases (Hagen et al. 1990). Experimentally, ablation of capsaicin-sensitive afferent neurons reduces the survival of a musculocutaneous flap (Kjartansson et al. 1987) and can lead to appearance of persistent skin wounds (Maggi et al. 1987a), aggravation of acid-induced skin lesions (Maggi et al. 1987a) and formation of keratitis-like lesions in the cornea (Fujita et al. 1984; Shimizu et al. 1984, 1987; Knyazev et al. 1990, 1991) of small rodents. A role of peptidergic afferent neurons in wound healing can also be envisaged from the ability of SP, NKA and CGRP to stimulate the proliferation of endothelial cells (Haegerstrand et al. 1990; Ziche et al. 1990), arterial smooth muscle cells (J. Nilsson et al. 1985, 1986; Payan 1985) and skin fibroblasts (J. Nilsson et al. 1985). These actions are likely to play a role in the angiogenesis of healing tissue, and SP is in fact able to induce neovascularization of the avascular rabbit cornea (Ziche et al. 1990).

Hyperactivity of afferent neurons may exacerbate and protract inflammation. The presence of inflammation is known to sensitize afferent nerve endings, a process in which mast cell- and leucocyte-derived inflammatory mediators such as histamine, prostaglandins and leukotrienes are likely to play a role (Kocher et al. 1987; Martin et al. 1987; Lynn 1988; McMahon and Koltzenburg 1990). Sensitization is expected to facilitate the neural release of SP and CGRP which, in turn, will enhance the liberation of histamine and other mediators from mast cells and leucocytes. Thus, there is a positive feedback loop which operates to cause hyperactivity of afferent nerve fibres and perpetuation of the inflammatory reaction (Lynn 1988). An additional aspect is lit up by the ability of SP and other sensory neuropeptides to control the immune system (see Payan et al. 1984; Stead et al. 1987a; G. Nilsson 1989; Donnerer et al. 1990; McGillis et al. 1990). It is, therefore, conceivable that sensitized afferent neurons enhance host defence reactions to injurious stimuli by a direct stimulant action on immunocompetent cells and by way of hyperaemia and increased vascular permeability which facilitate the delivery and accumulation of these cells in the inflamed tissue. In line with this proposal, antigen exposure has been found to increase nerve activity in afferent neurons (Undem et al. 1991), whilst the vascular reactions of the skin to allergen challenge in sensitized guinea pigs (Saria et al. 1983) and allergic humans (Lundblad et al. 1987; Wallengren and Möller 1988; McCusker et al. 1989; Wallengren 1991), and the vascular manifestations of acquired cold and heat urticaria (Tóth-Kása et al. 1983, 1984; G. Jancsó et al. 1985) are reduced after blockade of capsaicin-sensitive afferent nerve fibres or administration of an SP antagonist. Hyperreactive disorders of the skin such as psoriasis (Bernstein et al. 1986; Farber et al. 1986), bullous pemphigoid, eczema and photodermatoses (Wallengren et al. 1986) may also involve peptidergic afferent neurons. Further examples of chronic inflammation which may be driven by hyperactive afferent nerve fibres include vasomotor rhinitis (Marabini et al.

1988; Saria and Wolf 1988; Lacroix et al. 1991), asthma (Barnes et al. 1988) and chronic arthritis (Colpaert et al. 1983; Levine et al. 1984, 1986; Devillier et al. 1986b).

4 Afferent Neurons in Splanchnic Circulation

4.1 Innervation of the Splanchnic Vascular Bed by Peptidergic Afferent Neurons

The mesenteric arteries are innervated by SP- and CGRP-containing varicose and smooth nerve fibres which originate from dorsal root ganglia, and the superior mesenteric artery of the rat and guinea pig receives a particularly rich supply of peptidergic afferent nerve fibres that are sensitive to the neurotoxic action of capsaicin (Furness et al. 1982; Barja et al. 1983; Mulderry et al. 1985; Uddman et al. 1986; Wharton et al. 1986; Kawasaki et al. 1988, 1990a, b, c, 1991a; Wimalawansa and MacIntyre 1988; Fujimori et al. 1989, 1990; Del Bianco et al. 1991). The interlacing network of fibres is very dense in the main arteries and becomes loose towards the precapillary arterioles whilst the veins are sparsely innervated (Furness et al. 1982; Barja et al. 1983; Uddman et al. 1986).

In the wall of the gastrointestinal tract it is predominantly submucosal and mucosal blood vessels, especially arterioles, that are innervated by capsaicin-sensitive peptidergic afferent nerve fibres although other layers such as the myenteric plexus and circular muscle may also be supplied by these neurons (Furness et al. 1982; Minagawa et al. 1984; Papka et al. 1984; Sharkey et al. 1984; Gibbins et al. 1985, 1988; Rodrigo et al. 1985; Fehér et al. 1986; Uddman et al. 1986; Y. Lee et al. 1987b; Sternini et al. 1987, 1991b; Su et al. 1987; Galligan et al. 1988; Green and Dockray 1988; Varro et al. 1988; Chéry-Croze et al. 1989; Parkman et al. 1989; Kashiba et al. 1990). There are species differences in the neuropeptide expression of nerve fibres in the gastrointestinal tract (Brodin et al. 1983b; Keast et al. 1985; Sundler et al. 1991), and this is particularly well exemplified by an abundance of CGRP-containing afferent axons in the rat stomach (Green and Dockray 1988; Varro et al. 1988; Sundler et al. 1991) compared with a scarcity or absence of CGRP-containing fibres in the mucosa of the porcine and human stomach (Tsutsumi and Hara 1989; Sundler et al. 1991). Most, if not all, CGRP-containing nerve fibres in the rat stomach represent extrinsic afferent neurons that originate from dorsal root ganglia (Y. Lee et al. 1987b; Sternini et al. 1987; Su et al. 1987; Green and Dockray 1988; Varro et al. 1988; Kashiba et al. 1990; McGregor and Conlon 1991; Renzi et al. 1991). In contrast, vagal afferent nerve fibres contribute little to the SP and CGRP content of the rat stomach (Y. Lee et al. 1987b; Green and Dockray 1988; Varro et al. 1988). Vagal CGRP-containing afferent neurons, however, innervate the oesophagus (Rodrigo et al. 1985; Parkman et al. 1989).

The cell bodies of the spinal afferent neurons that contain CGRP and project to the gastrointestinal tract are of small to medium size, with diameters up to 45 μm (Su et al. 1987; Green and Dockray 1988; Kashiba et al. 1991), whereas the diameter of the SP-containing cell bodies hardly exceeds 35 μm (Green and Dockray 1988). SP- and CGRP-containing spinal afferent neurons also differ with regard to their axons. CGRP is present in 46% of unmyelinated, 33% of thinly myelinated and 17% of thickly myelinated axons (McCarthy and Lawson 1990), which is also true for CGRP-containing axons in the vagus nerve (Kakudo et al. 1988). In contrast, SP is found only in unmyelinated (50%) and thinly myelinated (20%) fibres (McCarthy and Lawson 1989). On their way to the gastrointestinal tract, the fibres of spinal afferent neurons pass through the prevertebral ganglia (Dalsgaard et al. 1983; Matthews and Cuello 1984; Y. Lee et al. 1987a; Su et al. 1987; Green and Dockray 1988; Lindh et al. 1988; Kashiba et al. 1990) where they give of collaterals that form axodendritic and axosomatic synapses with the principal sympathetic ganglion cells (Kondo and Yui 1981; M.R. Matthews and Cuello 1984; Y. Lee et al. 1987a; Green and Dockray 1988; Kondo and Yamamoto 1988; Lindh et al. 1988).

The peptidergic innervation of the gastrointestinal tract is complicated by the fact that there are two separate populations of peptide-containing neurons, extrinsic primary afferent and intrinsic enteric neurons, that have many neuropeptide transmitters in common (see Ekblad et al. 1985; Costa et al. 1986; Kirchgessner et al. 1988). This applies to the tachykinins SP and NKA, CGRP, VIP, somatostatin, opioid peptides and other neuronal markers. In addition, some markers of enteric and primary afferent neurons are also contained in parasympathetic (e.g. VIP, see Lundberg 1981) and sympathetic (e.g. somatostatin, see Lundberg et al. 1982; Costa and Furness 1984; Lindh et al. 1986, 1988; Vickers et al. 1990) efferent neurons. However, there are a number of discrete differences between these peptidergic neuron populations in terms of origin, projection, neurochemistry and function.

1. The chemical coding, i.e. the combination of peptides co-expressed in individual neurons, differs markedly among primary afferent, enteric and autonomic neurons (Costa et al. 1986). As in somatic afferent neurons, a population of the splanchnic CGRP-containing afferent neurons also contains SP (Gibbins et al. 1985, 1987a; Y. Lee et al. 1985; Rodrigo et al. 1985; Uddman et al. 1986; Su et al. 1987; Galligan et al. 1988; Green and Dockray 1988; Lindh et al. 1988; Chéry-Croze et al.

1989; Helke and Niederer 1990; Sundler et al. 1991). Conversely, SP and CGRP do not co-exist in enteric neurons of the rat and guinea pig (Gibbins et al. 1985; Costa et al. 1986; Belai and Burnstock 1987; Galligan et al. 1988; Goehler et al. 1988; Sundler et al. 1991).

2. The chemical identity of CGRP expressed by primary afferent and enteric neurons is different, at least in the rat. It seems as if most of the CGRP present in afferent neurons is CGRP-α (Mulderry et al. 1988; Varro et al. 1988; Noguchi et al. 1990; Sternini 1991), although some CGRP-β may also be expressed (Noguchi et al. 1990; Sternini 1991). In contrast, CGRP-β is the only form of CGRP present in enteric neurons (Mulderry et al. 1988; Sternini 1991).

3. Only the primary afferent neurons are sensitive to the excitatory and neurotoxic action of capsaicin whilst the autonomic (see Holzer 1991) and enteric (Barthó and Szolcsányi 1978; Szolcsányi and Barthó 1978; Holzer et al. 1980; Barthó et al. 1982; Furness et al. 1982; Donnerer et al. 1984; Holzer 1984; Papka et al. 1984; Sharkey et al. 1984; Kirchgessner et al. 1988; Geppetti et al. 1988; Green and Dockray 1988; Takaki and Nakayama 1989; Renzi et al. 1991) neurons are not sensitive to capsaicin.

4. Related to these different neurochemical and neuropharmacological characteristics is a diversity of physiological functions which are not detailed here except for the primary afferent neurons.

4.2 Blood Flow in Mesenteric Arteries

4.2.1 Local Regulation of Mesenteric Arterial Tone

4.2.1.1 Nature of the NANC Vasodilator Fibres

Functional evidence indicates that the mesenteric arteries are innervated not only by sympathetic vasoconstrictor fibres but also by NANC vasodilator fibres. The nature of the neuroeffector transmission in the inferior mesenteric artery of the guinea pig is governed by the frequency at which the periarterial nerve plexus is stimulated (Kreulen 1986; Hottenstein and Kreulen 1987). Whilst high-frequency (10–20 Hz) stimulation leads to noradrenergic depolarization and constriction, low-frequency (2–5 Hz) stimulation elicits slow inhibitory junction potentials and relaxation of the noradrenaline-precontracted arterial smooth muscle (Meehan et al. 1991). This hyperpolarization remains unaltered after removal of the endothelium or treatment of the preparation with α and β adrenoceptor antagonists, sympathetic neuron-blocking drugs, atropine, indomethacin or α,β-methylene ATP (Meehan et al. 1991). Electrical stimulation of isolated mesenteric arteries of the rat causes a similar dilatation of the vessels precontracted by α adrenoceptor agonists (Kawasaki et al. 1988, 1990a, b, c, 1991a; Axelsson et al. 1989; Fujimori et al. 1989, 1990; Han et al. 1990a, b; Li and Duckles 1992) or endothelin (Kawasaki et al. 1990b), especially when sympathetic nerve activity and noradrenergic transmission are blocked by 6-hydroxydopamine or guanethidine and prazosin. The vasodilator response depends on neural conduction but is independent of noradrenergic and cholinergic transmission and hence mediated by transmitters of NANC nerves (Kawasaki et al. 1988; Axelsson et al. 1989; Han et al. 1990b). Endothelium-derived factors (Axelsson et al. 1989; Han et al. 1990b; Li and Duckles 1992), histamine and prostaglandins (Kawasaki et al. 1988; Axelsson et al. 1989) have been ruled out as mediators. However, stimulation of the NANC vasodilator nerves appears to promote the production of vasoconstrictor prostaglandins (Li and Duckles 1991a).

The fibres responsible for NANC dilatation of mesenteric arteries in the dog (Rózsa et al. 1984, 1985), rat (Manzini and Perretti 1988; Fujimori et al. 1990; Han et al. 1990b; Kawasaki et al. 1990a, b, c, 1991a; Hottenstein et al. 1991) and guinea pig (Meehan et al. 1991) are sensitive to the stimulant and neurotoxic actions of capsaicin, whereas those in isolated bovine mesenteric arteries are insensitive to capsaicin (Axelsson et al. 1989). Stimulation of afferent nerve fibres by capsaicin mimics not only the hyperpolarizing response to NANC nerve stimulation (Meehan et al. 1991) but also the dilatation due to electrical stimulation of isolated mesenteric (Manzini and Perretti 1988; Fujimori et al. 1990; Meehan et al. 1991), hepatic and splenic (Bråtveit and Helle 1991) arteries of the rat and guinea pig. The capsaicin-evoked dilatation is independent of nerve impulse conduction and of cholinergic and noradrenergic transmission (Manzini and Perretti 1988). Periarterial administration of capsaicin to the superior mesenteric artery of the rat (Hottenstein et al. 1991) and dog (Rósza et al. 1984, 1985) in vivo augments blood flow in the that vessel, and the hyperaemia induced by injection of capsaicin into the superior mesenteric artery of the dog remains unaltered by blockade of adrenoceptor, dopamine and nicotinic acetylcholine receptors but is reduced by atropine (Rózsa et al. 1984, 1985). Capsaicin pretreatment of the rat and guinea pig isolated mesenteric arterial bed, to defunctionalize afferent nerve fibres, inhibits the vasodilator response to both capsaicin application (Manzini and Perretti 1988) and electrical nerve stimulation (Fujimori et al. 1990; Han et al. 1990b; Kawasaki et al. 1990a, b, c, 1991a; Li and Duckles 1991a; Meehan et al. 1991). A similar effect of capsaicin pretreatment is seen in the rat hepatic artery (Bråtveit and Helle 1991). These findings demonstrate that the NANC vasodilator fibres in the mesenteric/hepatic arteries of rat and dog arise from primary afferent neurons. The presence of peptides in, and their release from, these neurons as well as the inhibition of NANC vasodilatation by peptide antagonism indicates the peptidergic nature of the NANC vasodilator nerve fibres in the rat mesenteric arteries.

4.2.1.2 Release of Sensory Neuropeptides

Stimulation of periarterial nerves by electrical impulses or capsaicin application leads to a calcium-dependent release of CGRP (Fujimori et al. 1989, 1990; Del Bianco et al. 1991; Kawasaki et al. 1991a; Manzini et al. 1991). This stimulus-induced peptide release is inhibited after capsaicin-induced defunctionalization of sensory nerve fibres (Fujimori et al. 1990; Del Bianco et al. 1991; Manzini et al. 1991).

4.2.1.3 Neuropeptide Receptors on Mesenteric Blood Vessels

The abundance of CGRP in the superior mesenteric artery of the rat is matched by an abundance of CGRP binding sites (Wimalawansa and Mac-Intyre 1988). Rabbit mesenteric arteries bear a high number of VIP receptors that are coupled to adenylate cyclase (Huang and Rorstad 1987; Sidaway et al. 1989).

4.2.1.4 Vasodilator Activity of Sensory Neuropeptides

Among the sensory neuropeptides, CGRP is the most potent to dilate rat mesenteric arteries, but this effect is seen only upon local administration of the peptide whereas intravenous administration of CGRP to the conscious rat causes mesenteric vasoconstriction (Gardiner et al. 1989). In vitro, CGRP relaxes precontracted mesenteric arteries isolated from rat (Marshall et al. 1986; Kawasaki et al. 1988, 1990a, c; Manzini and Perretti 1988; Fujimori et al. 1989, 1990; Han et al. 1990a, b; Bråtveit et al. 1991; Li and Duckles 1991a, b, 1992), guinea pig (Uddman et al. 1986), rabbit (Nelson ct al. 1990) and humans (Törnebrandt et al. 1987). Like the NANC nerve-induced response, the vasodilatation in response to CGRP is independent of the endothelium (Li and Duckles 1992), and the time course of CGRP's effect is very similar to that of NANC nerve stimulation (Kawasaki et al. 1988, 1990a). Both in the rat and rabbit mesenteric vascular bed, rat α-CGRP is more potent as dilator than human α-CGRP (Marshall et al. 1986). Isolated bovine mesenteric arteries, though, appear to be insensitive to CGRP (Axelsson et al. 1989). Other vessels that are relaxed by CGRP in vitro include the left gastric artery from rat (Bråtveit et al. 1991) and guinea pig (Uddman et al. 1986), the rat splenic artery (Bråtveit and Helle 1991; Bråtveit et al. 1991), and the hepatic artery from rat (Bråtveit and Helle 1991; Bråtveit et al. 1991) and rabbit (Brizzolara and Burnstock 1991). The CGRP-induced relaxation of the hepatic and left gastric artery is independent of the endothelium (Bråtveit and Helle 1991; Bråtveit et al. 1991; Brizzolara and Burnstock 1991).

The effect of SP on mesenteric arteries varies with species and experimental conditions. SP is very weak in relaxing isolated bovine mesenteric arteries (Axelsson et al. 1989), and in the rat isolated mesenteric vascular bed SP, NKA and NKB fail to cause any effect (Barja et al. 1983; Kawasaki et al. 1988, 1990a; D'Orléans-Juste et al. 1991; Li and Duckles 1992). However, intravenous administration of SP to the rat (Rózsa and Jacobson 1989) and dog (Hallberg and Pernow 1975) increases mesenteric blood flow. A similar hyperaemia in the superior mesenteric artery of the dog (Melchiorri et al. 1977; Rósza et al. 1984, 1985) and pig (Schrauwen and Houvenaghel 1980) is seen after close arterial administration of SP. Provided that the endothelium is intact, isolated mesenteric arteries from the guinea pig are relaxed by SP (Bolton and Clapp 1986) as are human mesenteric artery preparations (Törnebrandt et al. 1987). The SP-induced relaxation of isolated mesenteric arteries from the rabbit is mediated by NK_1 receptors located on the endothelium (Zawadzki et al. 1981; Stewart-Lee and Burnstock 1989). Unlike arteries, rabbit isolated mesenteric veins are contracted by SP (Regoli et al. 1984), and blood flow in rat mesenteric venules is reduced by the peptide (B.J. Zimmermann et al. 1991). The tachykinin-induced contraction of isolated mesenteric veins from the rat is mediated by NK_3 receptors (D'Orléans-Juste et al. 1991).

Vasoactive intestinal polypeptide is able to dilate isolated mesenteric arteries from rat (Kawasaki et al. 1988) and humans (Törnebrandt et al. 1987), but VIP is less potent than CGRP (Kawasaki et al. 1988). Isolated bovine mesenteric arteries are only weakly relaxed by VIP (Axelsson et al. 1989). Intravascular administration of VIP to the rat stimulates mesenteric blood flow (Holliger et al. 1983; Rózsa and Jacobson 1989), an effect that also takes place after close arterial administration of VIP to the dog (Rózsa et al. 1985). Cholecystokinin-octapeptide is another dilator of mesenteric arteries of dog (Rózsa et al. 1985) and rat (Rózsa and Jacobson 1989). Somatostatin and methionine-enkephalin relax human mesenteric artery preparations precontracted with prostaglandin $F_{2\alpha}$ (Törnebrandt et al. 1987), whilst in the dog somatostatin reduces mesenteric blood flow (Rózsa et al. 1984, 1985).

4.2.1.5 Mediator Role of Sensory Neuropeptides in NANC Vasodilatation

A mediator role of peptides was first envisaged by the finding that α-chymotrypsin, a proteolytic enzyme, inhibits the capsaicin-induced relaxation of isolated mesenteric arteries from the rat (Manzini and Perretti 1988). Because desensitization to CGRP (Han et al. 1990a), the CGRP antagonist $CGRP_{8-37}$ (Han et al. 1990a; Kawasaki et al. 1991a) and a polyclonal CGRP antibody (Han et al. 1990b) are able to reduce mesenteric vasodilatation evoked by periarterial nerve stimulation, it would seem that CGRP is a major NANC vasodilator transmitter in the mesenteric vascular bed of the rat. Tachykinins do not seem to regulate the tone of isolated mesenteric arteries of the rat (Kawasaki et al. 1988) but appear to play a vasodilator role in the mesenteric arterial bed of the dog in vivo. Combined administration of antibodies to SP, VIP and cholecystokinin-octapeptide to the dog reduces the mesenteric vasodilator effect of capsaicin by 80%, the remaining vasodilatation being abolished by atropine (Rózsa et al. 1985).

4.2.1.6 Interaction Between the Peptidergic Vasodilator and Sympathetic Vasoconstrictor System

There is a mutual interaction between the NANC dilator and sympathetic constrictor neurons in the rat mesenteric arteries (Fig. 5). Low concentrations of noradrenaline, which are practically devoid of a direct action on the vascular tone, inhibit the dilatation of methoxamine-precontracted arteries in response to CGRP and electrical stimulation (Kawasaki et al. 1988). When endothelin, a peptide causing vasoconstriction independently of the sympathetic nervous system, is used to precontract the preparations, periarterial nerve stimulation fails to produce significant dilatation after blockade of the vasoconstrictor effects of endogenously released norad-

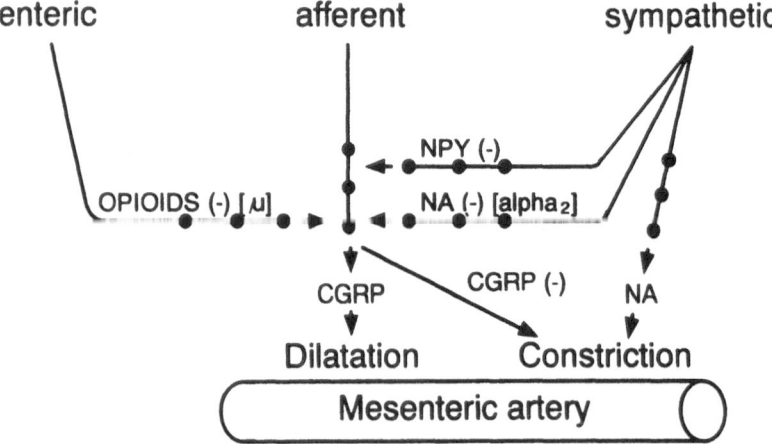

Fig. 5. Inhibitory action of sympathetic and possibly enteric neurons on the activity of afferent NANC vasodilator nerve fibres in mesenteric arteries. Sympathetic neurons can inhibit the release of the vasodilator peptide CGRP by way of prejunctional α_2 adrenoceptors and receptors for NPY. Prejunctional μ opioid receptors mediate the inhibitory effect of opioid peptides relased from enteric neurons. In addition, CGRP seems to inhibit the vasoconstrictor effect of noradrenaline (*NA*) by a postjunctional site of action. A possible participation of somatostatin is not shown

renaline with the α adrenoceptor antagonist prazosin (Kawasaki et al. 1990b). Vasodilatation is observed only when, in addition, sympathetic nerve activity and transmission are blocked by a combination of guanethidine and prazosin (Kawasaki et al. 1990b). Because concentrations of noradrenaline that do not alter vascular tone do not affect the vasodilator response to CGRP and acetylcholine (Kawasaki et al. 1991a), it seems as if endogenous noradrenaline, the release of which is prevented by guanethidine but not prazosin, inhibits NANC vasodilatation by a presynaptic site of action (Fig. 5), possibly by suppressing the release of the NANC vasodilator transmitters (Kawasaki et al. 1990b). This presynaptic action of noradrenaline, which is mediated by α_2 adrenoceptors (Kawasaki et al. 1990b), is shared by the sympathetic co-transmitter neuropeptide Y (NPY) (Fig. 5). Whereas the vasodilator effect of CGRP and acetylcholine is left unaltered, the NANC dilatation of methoxamine-precontracted mesenteric arteries is depressed by NPY (Kawasaki et al. 1991a; Li and Duckles 1991a). This and the inhibitory effect of NPY on the stimulus-induced release of CGRP from perimesenteric nerves indicate that NPY exerts a presynaptic inhibition of CGRP-containing vasodilator nerve endings in the rat mesenteric arteries (Kawasaki et al. 1991a).

Conversely, the NANC vasodilator nerves control the activity of sympathetic vasoconstrictor nerves. Defunctionalization of the vasodilator nerves by capsaicin pretreatment augments the vasoconstrictor effect of exogenous noradrenaline (Kawasaki et al. 1990a) and electrical stimulation of periarterial sympathetic neurons (Kawasaki et al. 1990a; Remak et al. 1990; Li and Duckles 1992) on isolated mesenteric arteries from the rat. Capsaicin pretreatment also augments sympathetically mediated contractions of the rat hepatic artery (Bråtveit and Helle 1991). In contrast, the release of endogenous noradrenaline is not altered by defunctionalization of afferent neurons (Han et al. 1990b; Kawasaki et al. 1990a). Since CGRP fails to alter the release of noradrenaline but inhibits the vasoconstrictor response to sympathetic nerve stimulation or exogenous noradrenaline, it would seem that the NANC vasodilator nerves inhibit noradrenaline-evoked vasoconstriction by a postjunctional site of action (Han et al. 1990b; Kawasaki et al. 1990a). This is consistent with the observation that noradrenergic excitatory junction potentials are reduced during inhibitory junction potentials caused by NANC nerve stimulation (Meehan et al. 1991).

These interactions between capsaicin-sensitive vasodilator nerves and sympathetic neurons are also seen in the rat in vivo. Periarterial nerve stimulation reduces blood flow in the superior mesenteric artery but, despite continued nerve stimulation, peak vasoconstriction is soon followed by a recovery of blood flow (Remak et al. 1990) and hyperaemia after cessation of nerve stimulation (Hottenstein et al. 1990). Both the initial autoregulatory espace from sympathetic vasoconstriction (Remak et al. 1990) and post-stimulation hyperaemia (Hottenstein et al. 1990) are inhibited by ablation of capsaicin-sensitive afferent neurons. Following defunctionalization of sympathetic neurons with reserpine, periarterial nerve stimulation causes vasodilatation, which is also blocked by ablation of sensory neurons (Remak et al. 1990). The sensory nerve-mediated autoregulatory escape response is the result of a peripheral interaction between vasodilator and vasoconstrictor nerve fibres (Remak et al. 1990). These observations re-affirm that the NANC vasodilator neurons control noradrenergic vasoconstrictor tone by a postjunctional site of action (Kawasaki et al. 1990a).

Another type of interaction between afferent and sympathetic efferent neurons is portrayed by the findings that permanent ablation of capsaicin-sensitive afferent neurons can lead to an increase in the transmitter content and/or innervation density of sympathetic nerve endings (Terenghi et al. 1986; Luthman et al. 1989), whereas long-term ablation of sympathetic neurons is followed by an increase in the peptidergic afferent innervation (Terenghi et al. 1986; Nielsch and Keen 1987; Aberdeen et al. 1990; Donnerer et al. 1991). These reciprocal alterations may arise from the competition of the two populations of neurons for nerve growth factor inasmuch as elimination of one population will enhance the availability of nerve growth factor for the surviving population of neurons and thus promote their development (Terenghi et al. 1986; Nielsch and Keen 1987; Luthman et al. 1989; Aberdeen et al. 1990). Although it has not yet been examined whether this interaction also takes place in the mesenteric arteries, there is one report to show that enteric NPY-containing neurons may also compensate in part for a long-term loss of sympathetic NPY-containing axons (Aberdeen et al. 1991). Reciprocal interactions between afferent and sympathetic neurons may be responsible for the decreased mesenteric blood flow which is seen in adult rats treated with capsaicin as neonates (Hottenstein et al. 1991), which could reflect hyperactivity of sympathetic vasoconstrictor neurons. Basal mesenteric arterial tone may thus be determined by a balance between NANC vasodilator and sympathetic vasoconstrictor activity. The observation that afferent neuron ablation in adult rats is followed by an increase in basal blood flow in the superior mesenteric artery (Hottenstein et al. 1991) is not yet explained, but it should be considered that capsaicin-sensitive afferent neurons could drive sympathetic efferent neurons and that after ablation of the afferent input sympathetic vasoconstrictor tone is diminished and hence blood flow increased.

4.2.1.7 Interaction with Opioid Peptide- and Somatostatin-Containing Neurons

Endogenous opioid peptides released by transmural electrical stimulation of rat mesenteric arteries inhibit NANC vasodilatation by activation of prejunctional µ opioid receptors (Li and Duckles 1991b) although pharmacological evidence indicates that prejunctional δ opioid receptors are also present (Li and Duckles 1991a). Because the mesenteric nerves of the guinea pig carry enkephalin- and dynorphin-containing fibres originating from the enteric nervous system (Lindh et al. 1988), it may be hypothesized that enteric neurons exert an inhibitory control over afferent nerve-mediated dilatation of mesenteric arteries (Fig. 5). The origin of somatostatin-containing nerve fibres in mesenteric nerves is complex, comprising afferent, sympathetic and enteric neurons (Lundberg et al. 1982; Costa and Furness 1984; Lindh et al. 1986, 1988; Vickers et al. 1990). The finding that somatostatin reduces mesenteric blood flow in the dog and diminishes the vasodilator effect of intraarterial capsaicin (Rózsa et al. 1984, 1985) may indicate that somatostatin modulates the vasodilator activity of peptidergic afferent nerve fibres. The ability of a somatostatin antibody to increase mesenteric blood flow by itself and to enhance the vasodilatation induced by capsaicin (Rózsa et al. 1985) supports this contention.

4.2.1.8 Physiological and Pathophysiological Implications

Vascular tone in the splanchnic bed is of significance for the whole cardiovascular system because the mesenteric resistance vessels are a primary target for the activity of sympathetic vasoconstrictor neurons (Greenway 1983; Jodal and Lundgren 1989). There is evidence that an imbalance between afferent vasodilator and sympathetic vasoconstrictor neurons may be of pathophysiological relevance for the etiology of hypertension. In spontaneously hypertensive rats the density of CGRP-containing nerve fibres around mesenteric arteries and the stimulus-induced release of CGRP from these fibres decrease with age when compared with that in normotensive Wistar-Kyoto rats (Kawasaki et al. 1990c, 1991b). These changes are associated with progressive hypertension, impairment of NANC dilatation, and accentuation of sympathetic constrictor responses in isolated mesenteric arteries from the rat (Kawasaki et al. 1990c, d, 1991b). The sensitivity of mesenteric arteries to the dilator action of CGRP, though, does not decrease but rather increases with age (Kawasaki et al. 1990c). It could be argued, therefore, that malfunction of peptidergic vasodilator neurons in mesenteric arteries is a factor in hypertension, vascular tone being augmented both by a deficit in neurogenic vasodilatation and by diminished inhibition of sympathetic vasoconstrictor tone.

4.2.2 Autonomic Reflex Regulation of Mesenteric Arterial Tone

In addition to local regulation of the vessel tone, capsaicin-sensitive afferent neurons also participate in the reflex regulation of mesenteric blood flow in response to stimuli administered to the mucosal or serosal surface of the intestine. Thus, mesenteric hyperaemia takes place when the sensory neuron stimulant capsaicin is administered into the small intestinal lumen of the dog (Rózsa et al. 1986) and rat (Rózsa and Jacobson 1989; Hottenstein et al. 1991) or when bile-oleate is given into the rat jejunum (Rózsa and Jacobson 1989). The responses to both capsaicin and bile-oleate are blocked by defunctionalization of capsaicin-sensitive afferent neurons and by local anaesthetics (Rózsa et al. 1986; Rósza and Jacobson 1989; Hottenstein et al. 1991), which indicates that they are the result of a neural reflex, the precise pathways of which are not known. Mesenteric hyperaemia due to intraluminal bile-oleate remains unaltered by blockade of muscarinic and nicotinic acetylcholine receptors, defunctionalization of noradrenergic neurons with reserpine, or antibodies to cholecystokinin-octapeptide and SP (Rósza and Jacobson 1989). While a participation of CGRP has not yet been tested, an antibody to VIP has been found to reduce the mesenteric vasodilatation evoked by intraluminal bile-oleate (Rózsa and Jacobson 1989). In delineating the neural pathways of this hyperaemic response, "axon reflexes" have been proposed to occur between intestinal and mesenteric collaterals of afferent neurons in response to chemical stimulation of afferent nerve endings in the gut (Rózsa et al. 1986; Rózsa and Jacobson 1989). This interpretation presupposes that VIP is a dilator transmitter of afferent nerve endings in the rat mesenteric arteries. However, although VIP may be expressed in some primary afferent neurons (G. Jancsó et al. 1981; Kuo et al. 1985; Skofitsch et al. 1985b), there is no evidence that the upper gastrointestinal tract of the rat is innervated by primary afferent neurons containing VIP (Green and Dockray 1988). In the guinea pig, mesenteric arteries are densely supplied by perivascular nerve fibres containing VIP (Della et al. 1983) which originate from enteric neurons sending intestinofugal axons along mesenteric arteries (Lindh et al. 1988; Vickers et al. 1990) because they remain unchanged after chemical ablation of capsaicin-sensitive afferent and noradrenergic efferent neurons (Della et al. 1983). Given that the situation is similar in the rat, it may be inferred that bile-oleate-induced mesenteric hyperaemia is due to a reflex in which both primary afferent (Rózsa and Jacobson 1989) and VIP-containing enteric neurons are involved. The possibility of this neural wiring is strengthened by the finding that primary afferent nerve fibres in the digestive tract can have synaptic contacts with enteric neurons (Neuhuber 1987).

Intestinal warming by mucosal or serosal application of fluid at 45°C gives rise to both constrictor and dilator responses in the superior mesenteric artery of the rat (Rózsa et al. 1988). These blood flow changes are abolished by intestinal application of a local anaesthetic and by defunctionalization of capsaicin-sensitive afferent neurons and seem to involve three different neural reflexes whose pathways run in the splanchnic but not vagal nerves (Rózsa et al. 1988). The decrease in mesenteric blood flow, which is the prevailing response, arises from two reflexes in which capsaicin-sensitive neurons form the afferent arc and sympathetic neurons the efferent vasoconstrictor arc, the difference being that one reflex has a spinal or supraspinal reflex centre, whilst the other reflex appears to be relayed in the coeliac/superior mesenteric ganglion complex (Rózsa et al. 1988). The hyperaemic response to intestinal warming, which is seen after defunctionalization of sympathetic neurons with reserpine, is relayed by pathways distal to the coeliac and superior mesenteric ganglion complex and, as cholinergic neurons are not involved, has been proposed to result from an axon reflex within collaterals of afferent neurons (Rózsa et al. 1988). The finding that systemic administration of an SP antibody inhibits both the vasoconstrictor and vasodilator reflex (Rózsa et al. 1988) suggests that SP serves as a transmitter between the afferent and efferent vasoconstrictor/vasodilator neurons. However, the inactivity of SP on isolated mesenteric arteries from the rat (Barja et al. 1983; Kawasaki et al. 1988, 1990a; D'Orléans-Juste et al. 1991) makes this peptide unlikely to be the vasodilator transmitter of the mesenteric hyperaemic response to intestinal warming. The possible participation of the mesenteric vasodilator peptides CGRP and VIP, and of enteric vasodilator neurons remains to be examined.

4.3 Blood Flow in the Gastrointestinal Mucosa

4.3.1 NANC Vasodilatation in the Gastrointestinal Mucosa

Electrical stimulation of the peripheral vagus nerve increases gastric mucosal blood flow, a response which in the cat (Martinson 1965) and rat (Morishita and Guth 1986) is in part mediated by cholinergic neurons and in part by NANC neurons. A NANC vasodilator response is also seen in the feline colon following electrical stimulation of the pelvic nerves (Hellström et al. 1991). The NANC vasodilator fibres in the rat vagus nerves are sensitive to the neurotoxic action of capsaicin and hence of primary afferent origin (Thiefin et al. 1990). The effect of vagus nerve stimulation is mimicked by application of the sensory neuron stimulant capsaicin to the rat stomach. Submucosal administration of capsaicin causes dilatation of submucosal arterioles (Chen et al. 1992), and intragastric administration of capsaicin enhances blood flow through the mucosa (Limlomwongse et al. 1979; Lippe et al. 1989b; Holzer et al. 1990a, 1991b; D.-S. Li et al. 1991; Matsumoto et al. 1991a, b; Takeuchi et al. 1991c; Leung 1992a). The hyperaemic effect of capsaicin in the rat gastric mucosa (Fig. 6) is mediated by afferent nerve fibres because it is absent in rats pretreated with a neurotoxic dose of capsaicin (Lippe et al. 1989b; Holzer et al. 1991b; Matsumoto et al. 1991a; Takeuchi et al. 1991c; Leung 1992a). These fibres, however, do not run in the vagus nerves but pass through the coeliac/superior mesenteric ganglion complex and most likely originate from spinal dorsal root ganglia (D.-S. Li et al. 1991). The involvement of neurons is also indicated by the ability of tetrodotoxin, a blocker of nerve conduction, to inhibit the gastric vasodilator response to intraluminal capsaicin (Holzer et al. 1991b), and the insensitivity of this effect to blockade of muscarinic acetylcholine receptors and of α and β adrenoceptors indicates the NANC nature of the participating neurons (Lippe et al. 1989b). Indomethacin has been reported to inhibit the vasodilator effect of capsaicin in the rat gastric mucosa (Takeuchi et al. 1991c), but it is not clear whether this result points to an involvement of vasodilator prostanoids (prostacyclin) because hyper-

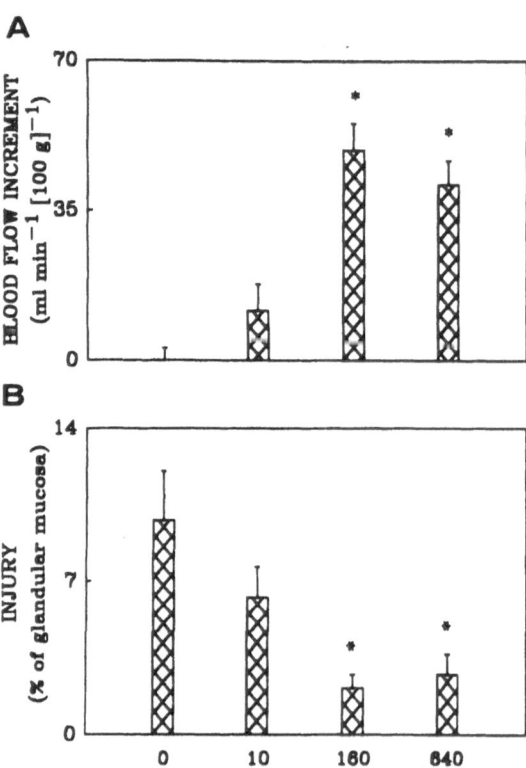

Fig. 6 A, B. Concentration-dependent effect of intragastric perfusion of capsaicin to induce mucosal hyperaemia (**A**) and to protect from gross mucosal damage induced by 25% ethanol (**B**) in the rat stomach. Means ± SEM, $n = 8$. *Asterisks, $p < 0.05$* versus controls (0 µmol l^{-1} capsaicin; U test). (Data from Holzer et al. 1991b)

aemia-inducing doses of capsaicin fail to alter the formation of prostanoids in the rat gastric mucosa (Holzer et al. 1990a).

Other regions in which capsaicin augments blood flow include the mucosa of the rat colon (Leung 1992b) and the mucosa and muscle of the rabbit oesophagus (Bass et al. 1991). As in the stomach, the capsaicin-induced hyperaemia in the rat colon (Leung 1992b) and rabbit oesophagus (Petersen et al. 1991b) depends on an intact sensory innervation of the tissue. In the rabbit oesophagus, however, both vagal and spinal afferent neurons participate in the hyperaemic response to capsaicin (Petersen et al. 1991b) which remains unaltered by indomethacin (Petersen et al. 1991a).

Afferent neurons are not only involved in the hyperaemic response to capsaicin but also play a role in the pathophysiological regulation of blood flow in the rat gastrointestinal mucosa. Thus, the hyperaemic response of the rat gastric mucosa to acid back-diffusion (Holzer et al. 1991a) or irritation with hypertonic saline (Matsumoto et al. 1991b) and the vasodilator response of the rat colonic mucosa to irritation with acetic acid (Leung 1992b) are mediated by capsaicin-sensitive afferent neurons. The increase in gastric mucosal blood flow evoked by intracisternal injection of a thyrotrophin-releasing hormone (TRH) analogue depends in part on capsaicin-sensitive afferent nerve fibres in the vagus nerves (Raybould et al. 1990).

4.3.2 Release of Sensory Neuropeptides in the Gastrointestinal Tract

Stimulation of peptidergic neurons in the digestive system leads to the release of bioactive peptides, but it is sometimes difficult to ascertain whether the peptides originate from intrinsic enteric, extrinsic afferent or extrinsic autonomic neurons. Thus, stimulation of the cat's vagus nerves results in the appearance of SP in the lumen of the stomach (Uvnäs-Wallensten 1978) and upper small intestine (Grönstad et al. 1985) and in increased plasma levels of SP in the portal vein blood (Grönstad et al. 1983). A test meal is similarly able to increase the concentrations of SP in the jejunal lumen and peripheral blood of the dog (Ferrara et al. 1987). Electrical stimulation of the pelvic nerves, mechanical stimulation of the anus, and distension of the rectum leads to a release of SP and NKA into the bloodstream of the cat (Hellström et al. 1991). The source of SP and NKA has not been determined in these experiments.

Capsaicin can be used to explore peptide release from sensory nerve fibres since this drug fails to release SP from enteric neurons of the guinea pig small intestine (Holzer 1984), and ablation of capsaicin-sensitive afferent neurons does not inhibit peristalsis-associated release of SP from enteric neurons (Donnerer et al. 1984). Intraarterial infusion of capsaicin causes release of SP into the vascular bed of the isolated guinea pig small intestine (Donnerer et al. 1984) and rat stomach (Kwok and McIntosh 1990).

Intravenous administration of capsaicin to the guinea pig in vivo is likewise followed by the appearance of SP in the gastric lumen, which is in keeping with the ability of capsaicin to release SP from the isolated guinea pig stomach (Renzi et al. 1988) and gallbladder (Maggi et al. 1989b). In contrast, capsaicin fails to release SP from the isolated rat stomach but increases the release of NKA (Renzi et al. 1991). The effect of capsaicin to induce tachykinin release is abolished after defunctionalization of capsaicin-sensitive neurons (Renzi et al. 1988, 1991; Maggi et al. 1989a, b; Geppetti et al. 1991).

Stimulation of primary afferent nerve endings by capsaicin is very effective in releasing CGRP from isolated tissues of the rat stomach (Geppetti et al. 1991; Ren et al. 1991; Renzi et al. 1991). Lowering the pH of the superfusing medium from 7.4 to 6 causes a similar release of CGRP (Geppetti et al. 1991), and the effects of both capsaicin and acidification are abolished by defunctionalization of capsaicin-sensitive primary afferent neurons. Peptone is also able to release CGRP from the isolated mucosa/submucosa of the rat gastric antrum (Manela et al. 1991). Intraarterial administration of capsaicin (Gray et al. 1989; Holzer et al. 1990b; Chiba et al. 1991), dibutyryl cyclic AMP or theophylline (Inui et al. 1989) leads to release of CGRP into the vascular bed of the perfused rat stomach. A capsaicin-induced release of CGRP from primary afferent neurons is also seen in the isolated gallbladder of the guinea pig (Maggi et al. 1989b) and in the muscle coat of the isolated rabbit colon (Mayer et al. 1990). In contrast, capsaicin fails to augment the release of CGRP from the isolated human small intestine but is able to evoke release of VIP (Maggi et al. 1989a). Electrical field stimulation and administration of caffeine augment the release of CGRP from the isolated rat ileum, whereas veratridine and potassium depolarization are without effect (Belai and Burnstock 1988).

4.3.3 Neuropeptide Receptors on Gastrointestinal Blood Vessels

Autoradiographically demonstrable binding sites for tachykinins are sparse on vessels of the digestive tract. Submucosal arterioles and venules in the human and canine gastrointestinal tract express only a low concentration of NK_1 receptor sites, whilst NK_2 and NK_3 binding sites are totally absent from gastrointestinal blood vessels (Gates et al. 1988; Mantyh et al. 1988a, b, 1989). A small number of NK_1 binding sites are also seen on large, but not small, blood vessels of the guinea pig ileum (Burcher and Bornstein 1988).

Conversely, CGRP receptors are densely distributed over blood vessels in the upper gastrointestinal tract of the dog. CGRP-α binding sites abound on medium-sized and small arteries and on arterioles in the submucosa and mucosa of the oesophagus, stomach, and duodenum, where they are expressed on both smooth muscle (tunica media) and endothelium (Gates et al. 1989). In contrast, CGRP-α binding sites are virtually absent from vessels in the jejunum and the more distal intestine of the dog. Unlike lymph nodules and lymph nodes, veins and venules do not exhibit binding sites for CGRP throughout the canine digestive system (Gates et al. 1989). Low levels of CGRP-binding sites are found in the submucosa of the human colon (Mantyh et al. 1989). In the rat high-affinity binding sites for CGRP occur throughout the gastrointestinal tract (Sternini et al. 1991a). They are especially numerous in the smooth muscle and endothelium of arteries and arterioles of the proximal small intestine (Sternini et al. 1991a).

Vasoactive intestinal polypeptide receptors are present in the small intestinal mucosa of the rat and rabbit (Sayadi et al. 1988) and on the smooth muscle of submucosal arterioles throughout the human gastrointestinal tract (R.P. Zimmerman et al. 1989).

4.3.4 Effects of Sensory Neuropeptides on Blood Flow in the Gastrointestinal Mucosa

The effects of SP on blood flow in the gastrointestinal tract are subject to species and regional differences. Intravenous administration of SP to the dog increases blood flow in the stomach (Yeo et al. 1982b) and small intestine (Burcher et al. 1977; Yeo et al. 1982b). Similarly, intraluminal application of SP stimulates mucosal but not muscular blood flow in the feline jejunum (Yeo et al. 1982a; Grönstad et al. 1983, 1986), an effect that does not involve neural conduction, histamine, prostaglandins, acetylcholine, noradrenaline or adrenaline (Grönstad et al. 1986). Close arterial infusion of NKA and SP to the feline colon leads to a hyperaemic response that is likewise independent of nerve conduction and cholinergic transmission (Hellström et al. 1991). The finding that NKA is more potent than SP in increasing blood flow in the cat's colon may suggest that NK_2 receptors mediate the vasodilator effect of tachykinins (Hellström et al. 1991). SP-induced arteriolar dilatation is also seen in the guinea pig isolated small intestine (Galligan et al. 1990). In contrast, SP and NKA fail to alter mucosal blood flow in the rat stomach (Yokotani and Fujiwara 1985; Holzer and Guth 1991), whilst SP can reduce gastric mucosal blood flow stimulated either by bethanechol or vagal nerve stimulation (Yokotani and Fujiwara 1985).

Intravenous administration of human CGRP-α and CGRP-β to anaesthetized rabbits potently stimulates mucosal blood flow in the stomach, CGRP-α being slightly more active than CGRP-β (Bauerfeind et al. 1989). The vasodilator acticity of CGRP in the stomach is considerably more pronounced than in the duodenum in which only CGRP-α, but not CGRP-β, causes mucosal hyperaemia (Bauerfeind et al. 1989). A similar regional heterogeneity is seen in the guinea pig in which rat CGRP-α is a potent dilator of the isolated gastroepiploic artery (Uddman et al. 1986) but fails to dilate submucosal arterioles in the small intestine (Galligan et al. 1990). Likewise, intravenous (DiPette et al. 1987; Lippe et al. 1989a; Bråtveit et al. 1991) or close arterial (Holzer and Guth 1991, D.-S. Li et al. 1991; Lippe and Holzer 1991) administration of rat CGRP-α potently augments mucosal blood flow in the rat stomach whilst blood flow in the small and large intestine is not altered (Bråtveit et al. 1991). The hyperaemic activity in the rat stomach is in keeping with the ability of topically applied CGRP ($0.1-10$ nmol l^{-1}) to dilate submucosal arterioles in this organ (Chen et al. 1992). The rise of mucosal blood flow induced by intraarterial CGRP-α is blocked by $CGRP_{8-37}$ (D.-S. Li et al. 1991) and by an inhibitor of NO synthesis (Lippe and Holzer 1991), which fits with an endothelium-dependent relaxant action of CGRP on precontracted arteries from the human stomach (Thom et al. 1987). Whether the vasodilator effect of CGRP in the digestive tract involves the release of secondary mediators such as prostacyclin (Crossman et al. 1987) and somatostatin (Dunning and Taborsky 1987; Helton et al. 1989) remains to be explored.

Vasoactive intestinal polypeptide is another neuropeptide that causes vasodilatation in the digestive system (Said and Mutt 1970). VIP-evoked relaxation of precontracted arteries from the human stomach requires an intact endothelium (Thom et al. 1987), and the dilator effect of VIP in submucosal arterioles of the guinea pig small intestine is attributed to the activation of enteric cholinergic vasodilator neurons (Galligan et al. 1990). Close arterial administration of VIP enhances mucosal blood flow in the canine stomach (Ito et al. 1988) and small intestine (Kachelhoffer et al. 1974) and in the stomach of the rat (Holzer and Guth 1991). Somatostatin, a further peptide present in afferent nerve fibres, is known to reduce gastrointestinal blood flow (see Jodal and Lundgren 1989) although the pertinent reports are not consistent with each other (see Leung and Guth 1985).

4.3.5 Mediator Role of Sensory Neuropeptides in Gastrointestinal Blood Flow

If sensory neuropeptides were to play a local role in mucosal vasodilatation in the rat stomach, CGRP would be the prime candidate owing to its high vasodilator activity in this organ (Lippe et al. 1989a; Holzer and Guth 1991; D.-S. Li et al. 1991; Lippe and Holzer 1991; Chen et al. 1992). Indeed, the dilator response of submucosal arterioles to topical capsaicin is inhibited by topical administration of a CGRP antagonist, $CGRP_{8-37}$, which indicates that the response is, at least in part, mediated by local release of CGRP (Chen et al. 1992). This inference is strengthened by the observation that close arterial administration of the CGRP antagonist to the rat stomach also attenuates the mucosal hyperaemia caused by intragastric capsaicin (D.-S. Li et al. 1991).

In the gastrointestinal tract of small rodents most, if not all, VIP is derived from enteric neurons intrinsic to the gut (Della et al. 1983; Ekblad et al. 1985; Costa et al. 1986), and it remains to be shown whether VIP released from capsaicin-sensitive neurons as shown in the human small intestine (Maggi et al. 1989a) plays a role in the regulation of local blood flow. The significance, if any, of the vascular actions of somatostatin for sensory nerve-mediated control of gastrointestinal blood flow is not known.

4.3.6 Pathways and Mediators of Gastric Mucosal Hyperaemia Due to Acid Back-Diffusion

Acid back-diffusion through a disrupted gastric mucosal barrier causes a prompt increase in gastric mucosal blood flow (Cheung et al. 1975; Ritchie 1975; Whittle 1977; Bruggeman et al. 1979; Starlinger et al. 1981; Oates 1990). The organization of the gastric circulation requires submucosal arterioles to be dilated in order to increase mucosal blood flow (Guth et al. 1989), which means that in the rat stomach the message of acid back-diffusion has to be transmitted over a 500-µm distance that separates the surface of the mucosa from submucosal arterioles. The finding that tetrodotoxin blocks the hyperaemic response to acid back-diffusion (Holzer et al. 1991a) indicates that communication between the acid-threatened surface mucosa and submucosal arterioles is accomplished by a neural pathway. Since defunctionalization of capsaicin-sensitive afferent neurons also blocks the response (Holzer et al. 1991a), it appears as if these neurons, which are particularly sensitive to H^+ ions (Bevan and Yeats 1991; Steen et al. 1992), monitor acid diffusing into the mucosal tissue. These chemosensitive nerve fibres originate from dorsal root ganglia (Raybould et al. 1992).

There are at least two different ways to explain the organization of the neural pathways and mediators of the hyperaemic response to acid back-diffusion, in which cholinergic vasodilator neurons do not participate (Bruggeman et al. 1979; Holzer et al. 1991a). One model holds that the increase in blood flow results from an axon reflex (Fig. 7) between mucosal and submucosal collaterals of afferent neurons (Holzer et al. 1991a; D.-S. Li et al. 1992). Diffusion of acid into the tissue excites the mucosal branches, after which an axon reflex enables nerve activity to be transmitted to branches around submucosal arterioles where release of peptides such as

CGRP will cause arteriolar dilatation (Fig. 7). Support for this concept comes from the findings that (a) acidification causes release of CGRP from rat gastric tissue (Geppetti et al. 1991); (b) CGRP is a potent vasodilator of arterioles in the rat gastric submucosa (Holzer and Guth 1991; D.-S. Li et al. 1991, 1992; Chen et al. 1992); (c) close arterial administration of the CGRP antagonist $CGRP_{8-37}$ to the rat stomach attenuates the mucosal hyperaemia caused by acid back-diffusion (D.-S. Li et al. 1992); and (d) the gastric mucosal vasodilator response to both CGRP (Lippe and Holzer 1991) and acid back-diffusion (Lippe and Holzer 1992) involves endothelium-derived NO.

However, this axon reflex model does not account for the finding that the rise of blood flow due to acid back-diffusion relies on intact conduction of nerve activity through afferent/efferent pathways in the splanchnic nerves (Fig. 8) and the coeliac/superior mesenteric ganglion complex (Holzer and Lippe 1992). As these pathways involve ganglionic transmission through nicotinic acetylcholine receptors (Holzer and Lippe 1992), it appears as if acid-induced hyperaemia is the result of a proper autonomic reflex in which the relay between afferent and efferent neurons takes place in the central nervous system and the efferent pathway consists of pre- and postganglionic neurons (Fig. 8). CGRP is likely to participate as a transmitter of the afferent neurons (Fig. 8) but, given its only localization in afferent nerve fibres of the rat stomach (Green and Dockray 1988), is unlikely to be the efferent vasodilator transmitter. The identity of this trans-

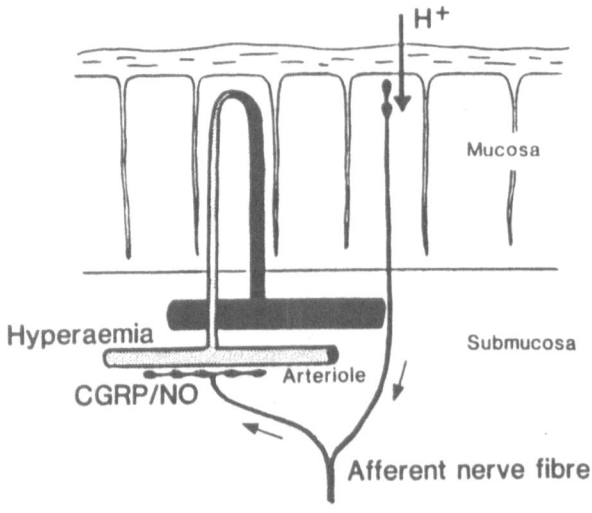

Fig. 7. Gastric mucosal hyperaemia due to acid back-diffusion as the result of an axon reflex between mucosal and submucosal collaterals of an afferent nerve fibre. Acid diffusing into the superficial mucosa activates the submucosal branch. Nerve activity is then transmitted to the submucosal branch which releases CGRP to activate the NO system causing arteriolar dilatation

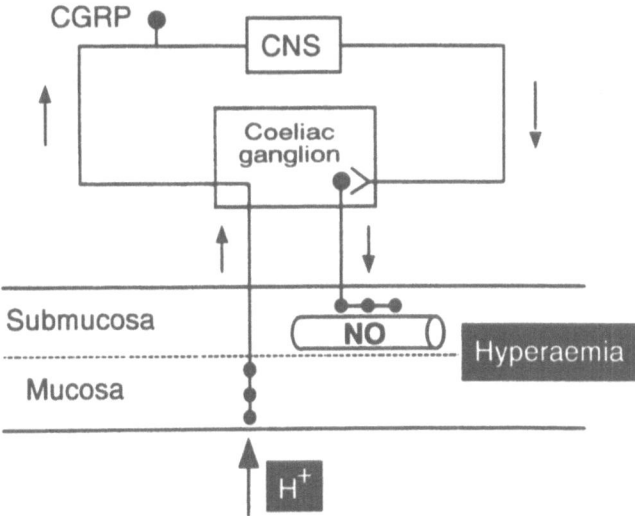

Fig. 8. Gastric mucosal hyperaemia due to acid back-diffusion as the result of an autonomic reflex involving afferent as well as pre- and postganglionic efferent neurons. CGRP is the transmitter of the afferent neurons, whilst the identity of the efferent vasodilator transmitter is not known except that it causes arteriolar dilatation via activation of the NO system

mitter which either is NO itself (see Moncada et al. 1991) or which causes vasodilatation via formation of NO (Lippe and Holzer 1992) remains to be determined. The finding that guanethidine pretreatment has no effect (Holzer and Lippe 1992) indicates that hyperaemia does not result from inhibition of noradrenergic sympathetic vasoconstrictor neurons.

The autonomic reflex model of acid-induced hyperaemia contrasts sharply with the axon reflex concept. The observation that acid back-diffusion increases blood flow not only in the gastric mucosa and submucosa but also in the coeliac and left gastric artery (P. Holzer, unpublished) indicates that hyperaemia is a response of the whole arterial tree supplying the stomach. This and a number of other points need to be considered in the further delineation of the neural pathways underlying the hyperaemic response to acid back-diffusion. First, the identity, projections and specific roles of all neurons involved in the hyperaemic reaction have to be determined. Special attention ought to be drawn to the possibility that enteric neurons projecting to the splanchnic blood vessels and the coeliac ganglion (Della et al. 1983; Lindh et al. 1988) participate in the acid-induced increase in gastric blood flow. Second, the nature of the neurotransmitters and the sites at which these transmitters communicate need to be identified. Thus, the site at which transmission via nicotinic acetylcholine receptors (Holzer and Lippe 1992) participates in the acid-induced hyperaemia is not clear, because ganglionic transmission may take place either in the coeliac ganglion or within the enteric nervous system. A similar uncertainty applies to the transmitter role of CGRP which may act as an afferent transmitter in the spinal cord, mediate afferent/sympathetic or afferent/enteric communication in the coeliac ganglion and/or act as a vasodilator transmitter in the gastric circulation (D.-S. Li et al. 1991, 1992). In addition, NO and/or VIP released from enteric neurons ought to be considered as possible vasodilator transmitters of the hyperaemic response to acid back-diffusion. Third, the role of the splanchnic nerves, transection of which abolishes the acid-induced hyperaemia (Holzer and Lippe 1992), needs to be defined. It is conceivable that the acid-induced increase in gastric blood flow is relayed by a peripheral neural circuitry which depends on an excitatory or inhibitory input from splanchnic nerve fibres.

Histamine H_1 receptors appear to play some role in the hyperaemic response to acid back-diffusion in the dog (Bruggeman et al. 1979) but not in the rat (Holzer et al. 1991a). Prostaglandins have been ruled out as mediators of gastric mucosal hyperaemia due to acid back-diffusion (Lippe and Holzer 1992), whilst the inhibitory effect of morphine may reflect the presence of opioid receptors on the neural pathways engaged in the vasodilator response (Holzer et al. 1991a).

4.3.7 Interaction with Other Vasodilator and with Vasoconstrictor Systems

There is indirect evidence that afferent nerve-dependent vasodilator mechanisms interact with other vasodilator and vasoconstrictor systems in the gastrointestinal mucosa. Endothelium-derived NO represents an important vasodilator system in the rat gastric mucosa, as inhibition of NO biosynthesis is followed by gastric mucosal vasoconstriction (Pique et al. 1989; Walder et al. 1990; Lippe and Holzer 1992). Afferent neurons and the NO system appear to exert a mutual control of mucosal blood flow because the vasoconstriction induced by blockade of NO synthesis is amplified after capsaicin-induced defunctionalization of afferent neurons (Tepperman and Whittle 1992).

Interactions between afferent neurons and sympathetic neurons have not yet been explored in the gastrointestinal mucosa except that defunctionalization of capsaicin-sensitive afferent neurons attenuates the autoregulatory escape reaction from adrenaline-induced vasoconstriction in the rat gastric mucosa (Leung 1992a), which suggests that vasoconstriction is counteracted by afferent nerve-mediated vasodilatation. A similar mode of interaction may be responsible for the finding that the vasoconstrictor effect of intravenous platelet-activating factor in the rat gastric mucosa is accentuated by prior defunctionalization of capsaicin-sensitive afferent neurons (Pique et al. 1990).

4.3.8 Cardiac and Blood Pressure Responses to Stimulation of Visceral Afferent Neurons

As in the skin and other tissues, capsaicin-sensitive afferent nerve fibres in the gastrointestinal tract participate in the autonomic reflex regulation of the cardiovascular system. Activation of visceral afferent fibres causes sympathetically mediated hypertension and tachycardia in the dog (Longhurst et al. 1980; Pitetti et al. 1988) and cat (Ordway and Longhurst 1983; Ordway et al. 1988). In the rat, activation of capsaicin-sensitive afferent nerve fibres by distension of the small intestine and trauma or chemical irritation of the peritoneum leads to an initial fall and a subsequent rise of blood pressure (Lembeck and Skofitsch 1982; Holzer et al. 1992). The initial fall of blood pressure involves noradrenergic sympathetic neurons (Holzer et al. 1992). The extent to which mesenteric and gastrointestinal circulation is affected by these cardiovascular reactions is not known.

4.3.9 Physiological and Pathophysiological Implications

4.3.9.1 Protection of Gastrointestinal Mucosa from Injury

Maintaining or increasing mucosal blood flow is thought to be a central element in the protection of the gastrointestinal mucosa from endogenous and exogenous injurious factors (see Guth et al. 1989; Oates 1990). Sensory nerve-mediated mucosal hyperaemia in the gastrointestinal tract is likely to play a similar role, and there is twofold experimental evidence for this contention.

First, ablation of capsaicin-sensitive afferent neurons causes aggravation of experimentally induced lesion formation in the gastric (Szolcsányi and Barthó 1981; Evangelista et al. 1986; Holzer and Sametz 1986; Esplugues et al. 1989; Esplugues and Whittle 1990; Szolcsányi 1990b; Chiba et al. 1991; Holzer et al. 1991a; Takeuchi et al. 1991a; Whittle and Lopez-Belmonte 1991; Raybould et al. 1992), duodenal (Maggi et al. 1987b; Takeuchi et al. 1991b) and colonic (Evangelista and Meli 1989; Eysselein et al. 1991b) mucosa.

Second, intragastric administration of capsaicin, to stimulate sensory nerve fibres, is able to protect against experimentally imposed lesions in the rabbit oesophageal (Bass et al. 1991), rat gastric (Szolcsányi and Barthó 1981; Holzer and Lippe 1988; Holzer et al. 1989, 1990a, 1991b; Robert et al. 1990; Szolcsányi 1990b; Williamson et al. 1990; Chiba et al. 1991; Peskar et al. 1991; Takeuchi et al. 1991c; Uchida et al. 1991) and rat distal colonic (Endoh and Leung 1990) mucosa. The protective effect of capsaicin is absent after ablation of capsaicin-sensitive afferent neurons (Holzer and Lippe 1988; Holzer et al. 1991b; Uchida et al. 1991). It needs to be considered, though, that the effects of sensory neuron manipulation on the protection of gastrointestinal mucosa depend on the type of injurious stimulus (Evangelista and Meli 1989; Robert et al. 1990) and gastrointestinal region (Holzer and Sametz 1986; Endoh and Leung 1990) unter study.

Several factors indicate a close relationship between afferent nerve-mediated mucosal hyperaemia and protection in the rat stomach.

1. There is a good correlation between the doses of capsaicin which increase blood flow and those which protect from ethanol injury (Fig. 6; Holzer et al. 1991b).
2. Both the protective (Chiba et al. 1991) and vasodilator (D.-S. Li et al. 1991) effect of intraluminal capsaicin on the rat gastric mucosa is blocked by the CGRP antagonist $CGRP_{8-37}$.
3. Like defunctionalization of capsaicin-sensitive afferent neurons, $CGRP_{8-37}$ significantly enhances the formation of mucosal damage caused by ethanol, indomethacin (Chiba et al. 1991) or acid back-diffusion (D.-S. Li et al. 1992). This finding is paralleled by exacerbation of ethanol-induced gastric mucosal damage in rats actively immunized against CGRP (Forster and Dockray 1991).
4. CGRP, which is a potent vasodilator in the rat stomach, also potently prevents experimental lesion formation in the stomach (Maggi et al. 1987c; Kolve and Taché 1989; Lippe et al. 1989a; Robert et al. 1990; Whittle and Lopez-Belmonte 1991).
5. The interaction between afferent neurons and the NO system in the control of gastric blood flow is matched by a similar interaction in gastric mucosal protection. Like capsaicin, NO is able to protect from gastric mucosal injury (MacNaughton et al. 1989), and blockade of NO biosynthesis inhibits the protective effect of capsaicin in the rat gastric mucosa (Peskar et al. 1991). Neither ablation of capsaicin-sensitive afferent neurons (Holzer and Sametz 1986; Holzer and Lippe 1988; Esplugues et al. 1989; Esplugues and Whittle 1990; Whittle et al. 1990; Whittle and Lopez Belmonte 1991) nor inhibition of NO synthesis (Hutcheson et al. 1990; Whittle et al. 1990) alone causes gastrointestinal mucosal damage. When, however, both vasodilator systems are eliminated, extensive damage develops in response to challenge of the gastric mucosa with acid (Whittle et al. 1990). A smiliar interaction exists between afferent neurons in the rat gastric mucosa and the vasoconstrictor endothelin-1 (Whittle and Lopez-Belmonte 1991).

In contrast to CGRP, the tachykinins SP and NKA fail to alter mucosal blood flow in the rat stomach, and subcutaneously administered SP is unable to prevent gastric damage (Evangelista et al. 1989; Robert et al. 1990), whilst intraperitoneal SP even enhances experimental gastric injury (Karmeli et al. 1991). The mechanism of the gastroprotective action of subcutaneously administered NKA (Evangelista et al. 1989) is not clear, but there is additional eivdence that tachykinins may be engaged in certain processes of gastric mucosal protection (Williamson et al. 1990).

The protective effect of afferent nerve stimulation by intraluminal capsaicin in the rat duodenum is associated with an increase in alkaline secretion (Takeuchi et al. 1991b), whilst a possible relationship to mucosal blood flow has not yet been determined.

To put these findings into proper pathophysiological perspective, it needs to be considered that capsaicin-sensitive afferent neurons represent probes that monitor a variety of potentially harmful chemicals including hydrochloric acid (Clarke and Davison 1978; Cervero and McRitchie 1982; Martling and Lundberg 1988; Forster et al. 1990; Bevan and Yeats 1991; Geppetti et al. 1991; Holzer et al. 1991a; Takeuchi et al. 1991b; Holzer and Lippe 1992; D.-S. Li et al. 1992; Raybould et al. 1992; Steen et al. 1992), acetic acid (Leung 1992b), hypertonicity (Forster et al. 1990; Matsumoto et al. 1991b; Tramontana et al. 1991), the bacterial peptide N-formyl-methionyl-leucyl-phenylalanine (Giuliani et al. 1991), platelet-activating factor (Rodrigue et al. 1988; McCusker et al. 1989; Pique et al. 1990; Sestini et al. 1990; Spina et al. 1991), endothelin-1 (Whittle and Lopez-Belmonte 1991) and other factors such as prostanoids, leukotrienes and bradykinin (see Maggi 1991; Rang et al. 1991). Recognition of these factors activates the neurons which, in turn, will call for mucosal hyperaemia and other reactions designed to prevent pending injury. Dysfunction of this neural emergency system is expected to weaken the resistance of the tissue to injurious stimuli and may thus be an aetiological factor in gastroduodenal ulcer disease.

4.3.9.2 Inflammatory Bowel Disease

Experimental colitis in the rat induced by trinitrobenzene sulphonic acid is worsened after ablation of capsaicin-sensitive afferent neurons, whereas colitis induced by ethanol or acetic acid remains unaffected (Evangelista and Meli 1989). Conversely, intraluminal capsaicin is able to reduce acetic acid-induced lesions in the distal, but not proximal, colon of the rat (Endoh and Leung 1990). A role of peptidergic neurons in colonic pathophysiology is further indicated by the findings that tissue levels of CGRP and SP in the rabbit colon decrease during the development of formaldehyde-immune complex colitis (Eysselein et al. 1991a) and that diminished CGRP concentrations are found in colonic muscle extracts of patients suffering from inflammatory bowel disease (Eysselein et al. 1991b). Inflammatory bowel disease is also associated with a dramatic up-regulation of SP receptors on submucosal arterioles and venules which normally express only a low concentration of SP receptor sites (Mantyh et al. 1988a, 1989). These findings suggest that SP, possibly released from sensory nerve endings, may be of relevance in the pathophysiology of the disease, although the precise functional implications of SP receptor up-regulation have not yet been determined. A widely accepted hypothesis for the pathogenesis of inflammatory bowel disease holds that bacterial or dietary antigens enter the tissue through an abnormally leaky epithelium (see Wallace 1990). Capsaicin-sensitive afferent neurons may play a role in spotting these threats and calling for appropriate protective measures as they are sensitive to the bacterial peptide *N*-formyl-methionyl-leucyl-phenylalanine (Giuliani et al. 1991) and the staphylococcal enterotoxin B (Alber et al. 1989), and participate in tissue reactions to allergens (Saria et al. 1983; Lundblad et al. 1987; Wallengren and Möller 1988; McCusker et al. 1989; Sestini et al. 1990).

4.4 Permeability in Splanchnic Blood Vessels

Intragastric capsaicin fails to alter vascular permeability in the rat stomach as examined by the Evans blue leakage technique (Holzer and Lippe 1988; Takeuchi et al. 1990). This observation is consistent with the finding that electrical stimulation of the peripheral vagus, splanchnic and pelvic nerves, respectively, fails to increase vascular permeability in the rat stomach, and small and large intestine down to the rectum (Lundberg et al. 1984a), although antidromic stimulation of lumbar dorsal roots produces plasma protein leakage in the distal large intestine (Szolcsányi 1988).

In contrast, protein extravasation is clearly seen in the oesophagus, bile system and mesentery, and there are sharp demarcation lines between the tissues in which electrical nerve stimulation elicits protein leakage and those in which is does not (Lundberg et al. 1984a). Defunctionalization of capsaicin-sensitive afferent neurons abolishes the exudative responses to electrical nerve stimulation in all tissues (Lundberg et al. 1984a). Activation of afferent nerve endings by intravenous administration of capsaicin increases vascular permeability only in those tissues in which electrical nerve stimulation is active (Saria et al. 1983; Lundberg et al. 1984a), except the duodenum in which some extravasation in response to intraluminal capsaicin is discernible (Maggi et al. 1987b).

A putative mediator role of SP in afferent nerve-mediated protein leakage can be deduced from the finding that intravenous administration of SP increases vascular permeability roughly in only those gastrointestinal tissues in which nerve stimulation by electrical impulses or intravenous capsaicin is also active (Saria et al. 1983; Lundberg et al. 1984a). Thus, the stomach, and small and large intestine down to the rectum do not respond to SP (Saria et al. 1983; Lundberg et al. 1984a; Maggi et al. 1987b), whilst

SP-induced plasma protein leakage is clearly seen in oesophagus, bile system, mesentery and anal mucosa (Saria et al. 1983; Lundberg et al. 1984a). SP also enhances the adhesion of leukocytes to the endothelium of rat mesenteric venules but does not stimulate their emigration (B.J. Zimmerman et al. 1991).

5 Summary and Open Questions

5.1 Control of Vessel Diameter and Permeability

The experimental data surveyed here identify a group of fine primary afferent neurons to be involved in the control of cutaneous and splanchnic microcirculation and thereby disclose a significant, new aspect for the neural regulation of tissue blood flow and vascular permeability. These data also signify an important target for the development of drugs and therapeutic strategies in the treatment of vascular dysfunction. Pharmacologically characterized by their sensitivity to capsaicin, these afferent neurons take part in the autonomic reflex control of vascular functions but in addition are able to directly regulate diameter and permeability of small blood vessels by way of release of vasoactive peptide mediators from their peripheral endings. In this function peptidergic afferent neurons are now considered in many tissues to be identical with NANC vasodilator neurons. Given their sensitivity to noxious stimuli, peptide release is induced by tissue irritation and/or trauma, and afferent nerve-mediated arteriolar dilatation and increase in venular permeability are the the prime elements of the inflammatory tissue reaction ("neurogenic inflammation"). It is not yet clear, however, whether the spatially separated processes of arteriolar dilatation and venular exudation are regulated by the same or different nerve fibres.

There is multiple evidence to implicate CGRP and SP as the principal mediators of perivascular NANC neurons.

1. These peptides are localized in afferent nerve endings supplying cutaneous and visceral blood vessels, and the bulk of peptides synthesized in the neuronal somata are transported to the peripheral nerve endings (Brimijoin et al. 1980; Keen et al. 1982).
2. Following stimulation, CGRP and SP are released from the peripheral nerve endings of afferent neurons.
3. These peptides are vasoactive substances. Vasodilatation is caused by both CGRP and SP, whilst vascular permeability is increased by SP only. However, there are positive and negative interactions between the vascular actions of CGRP, SP and other inflammatory mediators.

4. Immunoneutralization and peptide receptor antagonism attenuate the vasodilator and exudative responses to afferent nerve stimulation.

Afferent nerve-mediated control of vascular effector systems is not confined to skin and gastrointestinal tract but is operative in a variety of other tissues, notably the respiratory system (see Lundberg and Saria 1987; Barnes et al. 1988) and the cerebral vascular bed (see Sakas et al. 1989; Edvinsson et al. 1990).

5.2 Species and Tissue Diversity in the Local Vascular Functions of Peptidergic Afferent Neurons

Afferent nerve stimulation in the rat increases vascular permeability in the skin but fails to do so in the stomach and most regions of the small and large intestine, whereas vasodilatation is induced in both skin and splanchnic vascular bed. Hyperaemia is also seen in the porcine and human skin, whereas signs of increased vascular permeability (leakage of plasma proteins, oedema) in response to capsaicin-induced stimulation of afferent neurons are weak or absent. There are two principal explanations for this species and tissue diversity, differences in pre- and/or postjunctional mechanisms. Differences in prejunctional mechanisms are reflected by pathway-specific patterns of coexisting peptides in afferent neurons (Costa et al. 1986; Gibbins et al. 1987a, b; O'Brien et al. 1989) and by different chemical coding of afferent neurons in different species. As a consequence, the vascular responses to afferent nerve stimulation will vary according to the mixture of vasoactive peptides released from the nerve endings and the vascular actions caused by the peptides. The net vascular response will also be determined by interactions between the released peptides themselves. These considerations imply that identification of the participating transmitters and elucidation of the dynamics of co transmission will be indispensable for a full appreciation of the vascular control function of afferent neurons.

Species and tissue differences in postjunctional mechanisms relate to disparities in the effector systems. Different target tissues possess different densities of receptors for vasoactive peptides as is particularly evident in the gastrointestinal tract. The high vasodilator activity of CGRP in the stomach may be related to the high density of vascular CGRP receptors present in this region (Gates et al. 1989). Conversely, the inconsistent activity of tachykinins in increasing the diameter and permeability of gastrointestinal blood vessels is likely to be accounted for by the scarcity of autoradiographically demonstrable tachykinin binding sites on vessels of the digestive system (Burcher and Bornstein 1988; Gates et al. 1988;

Mantyh et al. 1988a, b, 1989). Other postjunctional differences, which have not yet been defined in sufficient detail, concern the existence of different peptide receptor types, different stimulus-effect coupling mechanisms, different secondary mediators and different peptide inactivation mechanisms. The available evidence indicates that it is primarily NK_1 and $CGRP_1$ receptors that mediate the direct effects of SP and CGRP on the vascular system, but this contention needs to be confirmed by the use of receptor-selective tachykinin and CGRP antagonists.

5.3 Interaction with Other Vasoactive Systems

The peptides released from afferent nerve terminals influence vascular effector systems both by an action on the vessels themselves and by interaction with other microvascular control systems. It is important to realize that the dynamics of this interactions change with time. The initial stages of the vascular responses to afferent nerve-derived peptides result primarily from a direct action on the vascular endothelium and muscle. Since there is no proof for the existence of specialized neuroeffector junctions, the peptides have to diffuse some distance before they reach their receptors on the effector cells. It has only recently been appreciated that the vascular endothelium and its autacoid NO play a very important role in setting vascular tone according to the physiological needs of the perfused tissue (Pohl 1990; Moncada et al. 1991). This endothelial vasodilator system appears to be an important target of afferent nerve-derived vasoactive peptides. Another vascular control system with which afferent vasodilator neurons are in mutual interaction are sympathetic vasoconstrictor neurons.

The more protracted stages of the vascular responses to afferent nerve-derived peptides involve additional intra- and extravascular processes. Extravascularly, SP and other sensory neuropeptides are capable of activating mast cells to release histamine and other factors. Intravascularly, leucocytes (particularly neutrophil granulocytes, but also monocytes and lymphocytes) are stimulated to adhere to the endothelium and to emigrate into the surrounding tissue. Furthermore, activated monocytes will release prostaglandin, thromboxane and cytokine mediators (Lotz et al. 1988; McGillis et al. 1990). Some of the mediators released from mast cells and monocytes are vasoactive by themselves and, in addition, can reactivate afferent nerve endings and thereby provide a positive feedback loop which reinforces the vascular actions of afferent nerve stimulation. Other factors such as opioid peptides released from immunocytes counteract this reinforcement of inflammatory processes.

Physiological functioning of the vascular system in the long term depends on the balanced interaction between all vascular control systems. Imbalances in these interactions, e.g. between the afferent vasodilator and sympathetic vasoconstrictor neurons or between afferent neurons and components of the immune system, are liable to cause pathological changes in the vascular system and in mechanisms depending on the vascular system.

5.4 Pathophysiological Relevance of Afferent Neurons in the Control of Vascular Effector Systems

5.4.1 Dual Role in the Maintenance of Homeostasis

The involvement of peptidergic afferent neurons in vascular effector control needs to be seen in context with the stimuli they respond to. These neurons are connected to chemoceptors, chemonociceptors and polymodal nociceptors which enable them to detect noxious stimuli that are potentially or actually harmful to the tissue. Whilst perception of noxious stimuli seems to be the primary task of fine afferent peptidergic neurons innervating somatic tissues, it is also innocuous, physiological stimuli including cholecystokinin, bile and distension (see Raybould and Taché 1988, 1989; Rózsa and Jacobson 1989; Esplugues et al. 1990; Forster et al. 1990; Jin and Nakayama 1990) that are among their modalities in visceral tissues. The overall function of these sensory neurons is to maintain homeostasis (see Prechtl and Powley 1990) which afferent vasoactive neurons accomplish by two different mechanisms (Fig. 1).

1. The local release of vasoactive peptide mediators leads to appropriate changes in the microcirculation at the very site of stimulation. Hyperaemia and increased vascular permeability facilitate the delivery of macromolecules and leucocytes to the tissue which, if the stimuli are noxious, will promote resistance of the tissue against further damage and aid the repair of injury.
2. The afferent function of the neurons transmits the perceived information to the central nervous system and initiates both voluntary and autonomic reflexes to maintain homeostasis (Fig. 1).

However, the local vascular and afferent functions of sensory neurons do not necessarily operate in parallel. The electrical stimulus frequencies required to elicit vasodilatation are considerably lower than those required to cause nociception (Kenins 1981; Szolcsányi 1984, 1988; Magerl et al. 1987; Brenan et al. 1988; Lynn 1988; Lisney and Bharali 1989). This demonstrates that the two functions of fine afferent neurons can be activated separately, and it remains to be determined whether they are mediated, at least in part, by different populations of neurons (see Jänig and Lisney 1989; Lisney and Bharali 1989). A systematic investigation of the stimulus modalities and intensities, which are required for activation of the local vascular and afferent functions of sensory neurons at the somatic and visceral level, is likely to shed light on this issue.

Related to these considerations is the question whether afferent neurons participate in the moment-to-moment control of vascular diameter and permeability or operate only when they are activated by adequate stimuli. Because the evidence for a continuous control of cutaneous (Sann et al. 1988) and mesenteric (Hottenstein et al. 1991) blood flow is only circumstantial and does not consider the complex interaction of afferent neurons with other microcirculation control systems, it appears as if afferent nerve-mediated vasodilatation reflects primarily a reaction to challenges of homeostasis. It is an ingenious design that those neurons which detect potential threats to the integrity of the tissue are per se able to initiate appropriate measures to cope with the danger on the spot. Vasoactive afferent neurons thus represent a system "of first line defense" (Lembeck 1983; Lembeck and Bucsics 1990) against trauma, a mechanism that was first embodied by the "nocifensor system" of Lewis (1937a, b). This function is portrayed by the ability of afferent nerve stimulation to protect from experimentally imposed damage in the skin (Maggi et al. 1987a) and gastrointestinal mucosa (Szolcsányi and Barthó 1981; Holzer and Lippe 1988; Holzer et al. 1989, 1990a, 1991b; Endoh and Leung 1990; Robert et al. 1990; Szolcsányi 1990b; Williamson et al. 1990; Bass et al. 1991; Chiba et al. 1991; Peskar et al. 1991; Takeuchi et al. 1991c; Uchida et al. 1991), processes which are very likely related to hyperaemia and increased vascular permeability.

Another phenomenon that is potentially relevant for homeostasis is the spread of cutaneous flare beyond the site at which the skin is irritated. This propagation of arteriolar dilatation may be considered as a measure to ensure that protective hyperaemia takes place not only in the challenged tissue but also in a "safety margin" (Holzer 1988). The spread of flare is most commonly explained as the result of an axon reflex between collaterals of fine afferent neurons (Fig. 3), but this concept has not yet been proven neurophysiologically and, in fact, has turned out very difficult to test experimentally. Uncritical extrapolation of the axon reflex concept to tissues other than the skin is a matter of "analogy speculation" which, in the absence of relevant neuroanatomical and neurophysiological data, leads to misconceptions as to how afferent neurons control the vascular system.

5.4.2 Acute Versus Chronic Neurogenic Inflammation

It is important to differentiate between the role of afferent neurons in acute inflammation and that in chronically inflamed tissue. Acute nociception, inflammation and activation of the immune system are important for protection against damage and for recovery from injury. Dysfunction of vasoactive afferent neurons is liable to cause homeostatic disorders and inadequate reactions to challenges of homeostasis. Exaggerated susceptibility to injurious factors is observed after defunctionalization of vasoactive afferent neurons in the skin and gastrointestinal mucosa, and malfunction of the NANC vasodilator system in the mesenteric arteries seems to be a factor in the aetiology of hypertension (Kawasaki et al. 1990c, d, 1991b)

The role of vasoactive afferent neurons seems to be quite different in chronic inflammation. In this situation, peptidergic afferent neurons can become hyperreactive and contribute to hyperalgesia and perpetuation of the inflammatory and immune reactions to an initial tissue insult. Such an adverse role of vasoactive afferent neurons appears to be involved not only in certain hyperreactive disorders of the skin but also in vascular headaches (see Moskowitz et al. 1989; Olesen and Edvinsson 1991), chronic obstructive airway disorders (see Barnes et al. 1988) and in chronic arthritis (see S.M. Louis et al. 1990; Basbaum and Levine 1991). Two factors may be instrumental in this respect: (a) some of the inflammatory mediators (e.g. prostaglandins, leukotrienes, platelet-activating factor, histamine, bradykinin), that trauma and afferent nerve-derived peptides release from mast cells, leucocytes and other cellular systems, are able to reactivate sensory nerve endings and thus provide a continuous drive to maintain inflammation; (b) inflammatory mediators such as prostaglandins are able to sensitize nociceptive neurons which will further up-regulate afferent nerve-mediated inflammatory and immune reactions.

An important point to consider in this respect is the existence of "silent nociceptors" in skin, joints and visceral tissues (see Schaible and Schmidt 1988; McMahon and Koltzenburg 1990). This term describes unmyelinated afferent nerve fibres (C fibres) which in the normal tissue fail to respond to mechanical and thermal noxious stimuli although they are responsive to algesic chemicals. In the inflamed tissue, however, silent nociceptors become sensitized to mechanical stimuli and seem to be responsible for the hyperalgesia accompanying chronic inflammation (McMahon and Koltzenburg 1990). Although these hyperreactive neurons are sensitive to the neurotoxic action of capsaicin (Barthó et al. 1990b), it remains to be determined whether they are also active in the control of local vascular functions and whether they contribute to the perpetuation of the inflammatory reaction.

Acknowledgements. I am very grateful to Drs. Carlo A. Maggi, Irmgard T. Lippe and Josef Donnerer for their helpful comments on the manuscript. Dr. Ulrike Holzer-Petsche prepared the computer-generated graphs, Irmgard Russa helped in typing the manuscript and Wolfgang Schluet was of administrative assistance. Research done in the author's laboratory was supported by the Austrian Scientific Research Council (grants 4641, 5552 and 7845).

References

Aberdeen J, Corr L, Milner P, Lincoln J, Burnstock G (1990) Marked increases in calcitonin gene-related peptide-containing nerves in the developing rat following long-term sympathectomy with guanethidine. Neuroscience 35:175–184

Aberdeen J, Moffitt D, Burnstock G (1991) Increases in NPY in non-sympathetic nerve fibres supplying rat mesenteric vessels after immunosympathectomy. Regul Pept 34:43–54

Alber G, Scheuber PH, Reck B, Sailer-Kramer B, Hartmann A, Hammer DK (1989) Role of substance P in immediate-type skin reactions induced by staphylococcal enterotoxin B in unsensitized monkeys. J Allergy Clin Immunol 84:880–885

Ali H, Leung KB, Pearce FL, Hayes NA, Foreman JC (1986) Comparison of the histamine-releasing action of substance P on mast cells and basophils from different species and tissues. Int Arch Allergy Appl Immunol 79:413–418

Altura BM, Gebrewold A, Burton RW (1985) Failure of microscopic metarterioles to elicit vasodilator responses to acetylcholine, bradykinin, histamine and substance P after ischemic shock, endotoxemia, and trauma: possible role of endothelial cells. Microcirc Endothelium Lymphatics 2:121–127

Alving K, Matran R, Lundberg JM (1991a) Capsaicin-induced local effector responses, autonomic reflexes and sensory neuropeptide depletion in the pig. Naunyn Schmiedebergs Arch Pharmacol 343:37–45

Alving K, Sundström C, Matran R, Panula P, Hökfelt T, Lundberg JM (1991b) Association between histamine-containing mast cells and sensory nerves in the skin and airways of control and capsaicin-treated pigs. Cell Tissue Res 264:529–538

Amann R, Donnerer J, Lembeck F (1990) Activation of primary afferent neurons by thermal stimulation. Influence of ruthenium red. Naunyn Schmiedebergs Arch Pharmacol 341:108–113

Anand P, Bloom SR, McGregor GP (1983) Topical capsaicin pretreatment inhibits axon reflex vasodilatation caused by somatostatin and vasoactive intestinal polypeptide in human skin. Br J Pharmacol 78:665–669

Ando K, Ito Y, Ogata E, Fujita T (1992) Vasodilating actions of calcitonin gene-related peptide in normal man: comparison with atrial natriuretic peptide. Am Heart J 123: 111–116

Andrews PV, Helme RD (1989) Tachykinin-induced vasodilatation in rat skin measured with a laser-Doppler flowmeter: evidence for receptor-mediated effects. Regul Pept 25: 267–275

Andrews PV, Helme RD, Thomas KL (1989) NK-1 receptor mediation of neurogenic plasma extravasation in rat skin. Br J Pharmacol 97:1232–1238

Aronin N, Leeman SE, Clements RS (1987) Diminished flare response in neuropathic diabetic patients. Comparison of effects of substance P, histamine, and capsaicin. Diabetes 36:1139–1143

Arvier PT, Chahl LA, Ladd RJ (1977) Modification by capsaicin and compound 48/80 of dye leakage induced by irritants in the rat. Br J Pharmacol 59:61–68

Axelsson KL, Ljusegren ME, Ahlner J, Grundström N (1989) A novel neurogenic vasodilator mechanism in bovine mesenteric artery. Circ Res 65:903–908

Barja F, Mathison R, Huggel H (1983) Substance P-containing nerve fibres in large peripheral blood vessels of the rat. Cell Tissue Res 229:411–422

Barnes PJ, Brown MJ, Dollery CT, Fuller RW, Heavey DJ, Ind PW (1986) Histamine is released from skin by substance P but does not act as the final vasodilator in the axon reflex. Br J Pharmacol 88:741–745

Barnes PJ, Chung KF, Page CP (1988) Inflammatory mediators and asthma. Pharmacol Rev 40:49–84

Barthó L, Szolcsányi J (1978) The site of action of capsaicin on the guinea-pig isolated ileum. Naunyn Schmiedebergs Arch Pharmacol 305:75–81

Barthó L, Szolcsányi J (1981) Opiate agonists inhibit neurogenic plasma extravasation in the rat. Eur J Pharmacol 73:101–104

Barthó L, Sebök B, Szolcsányi J (1982) Indirect evidence for the inhibition of enteric substance P neurones by opiate agonists but not by capsaicin. Eur J Pharmacol 77:273–279

Barthó L, Ernst R, Pierau F-K, Sann H (1990a) Opioid effects on electrically-induced vasodilatation and on flare reactions in the pig's skin. Eur J Pharmacol 183:1293–1294

Barthó L, Stein C, Herz A (1990b) Involvement of capsaicin-sensitive neurones in hyperalgesia and enhanced opioid antinociception in inflammation. Naunyn Schmiedebergs Arch Pharmacol 342:666–670

Basbaum AI, Levine JD (1991) The contribution of the nervous system to inflammation and inflammatory disease. Can J Physiol Pharmacol 69:647–651

Bass BL, Trad KS, Harmon JW, Hakki FZ (1991) Capsaicin-sensitive nerves mediate esophageal mucosal protection. Surgery 110:419–426

Bauerfeind P, Hof R, Cucala M, Siegrist S, von Ritter C, Fischer JA, Blum AL (1989) Effects of hCGRP I and II on gastric blood flow and acid secretion in anesthetized rabbits. Am J Physiol 256:G145–G149

Bayliss WM (1901) On the origin from the spinal cord of the vaso-dilator fibres of the hind-limb, and on the nature of these fibres. J Physiol (Lond) 26:173–209

Beglinger C, Born W, Münch R, Kurtz A, Gutzwiller J-P, Jäger K, Fischer JA (1991) Distinct hemodynamic and gastric effects of human CGRP I and II in man. Peptides 12: 1347–1351

Belai A, Burnstock G (1987) Selective damage of intrinsic calcitonin gene-related peptidelike immunoreactive enteric nerve fibers in streptozotocin-induced diabetic rats. Gastroenterology 92:730–734

Belai A, Burnstock G (1988) Release of calcitonin gene-related peptide from rat enteric nerves is Ca^{2+}-dependent but is not induced by K^+ depolarization. Regul Pept 23: 227–235

Benarroch EE, Low PA (1991) The acetylcholine-induced flare response in evaluation of small fiber dysfunction. Ann Neurol 29:590–595

Benjamin N, Dollery CT, Fuller RW, Larkin S, McEwan SE (1987) The effects of calcitonin gene-related peptide and substance P on resistance and capacitance vessels. Br J Pharmacol 90:39P

Bény JL, Brunet PC, Huggel H (1989) Effects of substance P, calcitonin gene-related peptide and capsaicin on tension and membrane potential of pig coronary artery in vitro. Regul Pept 25:25–36

Benyon CR, Robinson C, Church MK (1989) Differential release of histamine and eicosanoids from human skin mast cells activated by IgE-dependent and non-immunological stimuli. Br J Pharmacol 97:898–904

Beresford IJM, Birch BJ, Hagan RM, Ireland SJ (1991) Investigation into species variants in tachykinin NK_1 receptors by use of the non-peptide antagonist, CP-96,345. Br J Pharmacol 104:292–293

Bernstein JE, Swift RM, Soltani K, Lorincz AL (1981) Inhibition of axon reflex vasodilatation by topically applied capsaicin. J Invest Dermatol 76:394–395

Bernstein JE, Parish RC, Rapaport M, Rosenbaum MM, Roenigk HH (1986) Effect of topically applied capsaicin on moderate and severe psoriasis vulgaris. J Am Acad Dermatol 15:504–507

Bevan S, Szolcsányi J (1990) Sensory neuron-specific actions of capsaicin – mechanisms and applications. Trends Pharmacol Sci 11:330–333

Bevan S, Yeats J (1991) Protons activate a cation conductance in a sub-population of rat dorsal root ganglion neurones. J Physiol (Lond) 443:145–161

Bharali LAM, Lisney SJW (1988) Reinnervation of skin by polymodal nociceptors in rats. Prog Brain Res 74:247–251

Bjerring P, Arendt-Nielsen L (1990) Inhibition of histamine skin flare reaction following repeated topical applications of capsaicin. Allergy 45:121–125

Björklund H, Dalsgaard CJ, Jonsson CE, Hermansson A (1986) Sensory and autonomic innervation of non-hairy and hairy human skin. An immunohistochemical study. Cell Tissue Res 243:51–57

Blumberg H, Wallin BG (1987) Direct evidence of neurally mediated vasodilatation in hairy skin of the human foot. J Physiol (Lond) 382:105–121

Blumberg S, Teichberg VI, Charli JL, Hersh LB, McKelvy JF (1980) Cleavage of substance P to an N-terminal tetrapeptide and a C-terminal heptapeptide by a post-proline cleaving enzyme from bovine brain. Brain Res 192:477–486

Böckers M, Benes P, Bork K (1989) Persistent skin ulcers, mutilations, and acro-osteolysis in hereditary sensory and autonomic neuropathy with phospholipid excretion. J Am Acad Dermatol 21:736–739

Bolton TB, Clapp LH (1986) Endothelial-dependent relaxant actions of carbachol and substance P in arterial smooth muscle. Br J Pharmacol 87:713–723

Bond SM, Cervero F, McQueen DS (1982) Influence of neonatally administered capsaicin on baroreceptor and chemoreceptor reflexes in the adult rat. Br J Pharmacol 77:517–524

Brain SD, Williams TJ (1985) Inflammatory oedema induced by synergism between calcitonin gene-related peptide (CGRP) and mediators of increased vascular permeability. Br J Pharmacol 86:855–860

Brain SD, Williams TJ (1988) Substance P regulates the vasodilator activity of calcitonin gene-related peptide. Nature 335:73–75

Brain SD, Williams TJ (1989) Interactions between the tachykinins and calcitonin gene-related peptide lead to the modulation of oedema formation and blood flow in rat skin. Br J Pharmacol 97:77–82

Brain SD, Williams TJ, Tippins JR, Morris HR, MacIntyre I (1985) Calcitonin gene-related peptide is a potent vasodilator. Nature 313:54–56

Brain SD, Tippins JR, Morris HR, MacIntyre I, Williams TJ (1986) Potent vasodilator activity of calcitonin gene-related peptide in human skin. J Invest Dermatol 87:533–536

Brain SD, Petty RG, Lewis JD, Williams TJ (1990) Cutaneous blood flow responses in the forearms of Raynaud's patients induced by local cooling and intradermal injections of CGRP and histamine. Br J Clin Pharmacol 30:853–859

Bråtveit M, Helle KB (1991) Vasodilation by calcitonin gene-related peptide (CGRP) and by transmural stimulation of the methoxamine-contracted rat hepatic artery after pretreatment with guanethidine. Scand J Clin Lab Invest 51:395–402

Bråtveit M, Haugan A, Helle KB (1991) Effects of calcitonin gene-related peptide (CGRP) on regional haemodynamics and on selected hepato-splanchnic arteries from the rat: a comparison with VIP and atriopeptin II. Scand J Clin Lab Invest 51:167–174

Brenan A, Jones L, Owen NR (1988) The demonstration of the cutaneous distribution of saphenous nerve C-fibres using a plasma extravasation technique in the normal rat and following nerve injury. J Anat 157:57–66

Brimijoin S, Lundberg JM, Brodin E, Hökfelt T, Nilsson G (1980) Axonal transport of substance P in the vagus and sciatic nerves of the guinea pig. Brain Res 191:443–457

Brizzolara AL, Burnstock G (1991) Endothelium-dependent and endothelium-independent vasodilatation of the hepatic artery of the rabbit. Br J Pharmacol 103:1206–1212

Brodin E, Gazelius B, Lundberg JM, Olgart L (1981) Susbtance P in trigeminal nerve endings: occurrence and release. Acta Physiol Scand 111:501–503

Brodin E, Gazelius B, Panopoulos P, Olgart L (1983a) Morphine inhibits substance P release from peripheral sensory nerve endings. Acta Physiol Scand 117:567–570

Brodin E, Sjölund K, Håkanson R, Sundler F (1983b) Substance P-containing nerve fibres are numerous in human but not in feline intestinal mucosa. Gastroenterology 85:557-564

Bruce NA (1910) Über die Beziehung der sensiblen Nervenendigungen zum Entzündungsvorgang. Arch exp Pathol Pharmakol 63:424-433

Bruggeman TM, Wood JG, Davenport HW (1979) Local control of blood flow in the dog's stomach: vasodilatation caused by acid back-diffusion following topical application of salicylic acid. Gastroenterology 77:736-744

Buck SH, Burks TF (1986) The neuropharmacology of capsaicin – review of some recent observations. Pharmacol Rev 38:179-226

Buck SH, Harbeson SL, Hassmann CF, Shatzer SA, Rouissi N, Nantel F, Van Giersbergen PLM (1990) [Leu^{9}ψ(CH$_2$NH)-Leu^{10}]-neurokinin A(4-10) (MDL 28,564) distinguishes tissue tachykinin peptide NK$_2$ receptors. Life Sci 47:PL37-PL41

Buckley TL, Brain SD, Williams TJ (1990) Ruthenium red selectively inhibits oedema formation and increased blood flow induced by capsaicin in rabbit skin. Br J Pharmacol 99:7-8

Buckley TL, Brain SD, Rampart M, Williams TJ (1991) Time-dependent synergistic interactions between the vasodilator neuropeptide, calcitonin gene-related peptide (CGRP) and mediators of inflammation. Br J Pharmacol 103:1515-1519

Buckley TL, Brain SD, Jose PJ, Williams TJ (1992) The partial inhibition of inflammatory responses induced by capsaicin using the Fab fragment of a selective calcitonin gene-related peptide antiserum in rabbit skin. Neuroscience 48:963-968

Bucsics A, Holzer P, Lembeck F (1983) The substance P content of peripheral tissues in several mammals. Peptides 4:451-455

Bueb JL, Mousli M, Bronner C, Rouot B, Landry Y (1990) Activation of G$_i$-like proteins, a receptor-independent effect of kinins in mast cells. Mol Pharmacol 38:816-822

Bunnett NW, Orloff MS, Turner AJ (1985) Catabolism of substance P in the stomach wall of the rat. Life Sci 37:599-606

Burcher E, Bornstein JC (1988) Localization of substance P binding sites in submucous plexus of guinea pig ileum, using whole-mount autoradiography. Synapse 2:232-239

Burcher E, Atterhög JH, Pernow B, Rosell S (1977) Cardiovascular effects of substance P: effects on the heart and regional blood flow in the dog. In: von Euler US, Pernow P (eds) Substance P. Raven, New York, pp 261-268

Carpenter SE, Lynn B (1981) Vascular and sensory responses of human skin to mild injury after topical treatment with capsaicin. Br J Pharmacol 73:755-758

Carter RB, Francis WR (1991) Capsaicin desensitization to plasma extravasation evoked by antidromic C-fiber stimulation is not associated with antinociception in the rat. Neurosci Lett 127:43-45

Celander O, Folkow B (1953a) The nature and the distribution of afferent fibres provided with the axon reflex arrangement. Acta Physiol Scand 29:359-370

Celander O, Folkow B (1953b) The correlation between the stimulation frequency and the dilator response evoked by 'antidromic' excitation of the thin afferent fibres in the dorsal roots. Acta Physiol Scand 29:371-376

Cervero F, McRitchie HA (1982) Neonatal capsaicin does not affect unmyelinated efferent fibers of the autonomic nervous system: functional evidence. Brain Res 239:283-288

Chahl LA (1979) The effect of putative peptide neurotransmitters on cutaneous vascular permeability in the rat. Naunyn Schmiedebergs Arch Pharmacol 309:159-163

Chahl LA (1988) Antidromic vasodilatation and neurogenic inflammation. Pharmacol Ther 37:275-300

Chahl LA, Chahl JS (1986) Plasma extravasation induced by dynorphin-(1-13) in rat skin. Eur J Pharmacol 124:343-347

Chahl LA, Manley SW (1980) Inflammatory peptide in spinal cord: evidence that the mediator of antidromic vasodilatation is not substance P. Neurosci Lett 18:99-103

Chahl LA, Lynch AM, Thornton CA (1984) Effect of substance P antagonists on plasma extravasation induced by substance P and capsaicin in rat skin. In: Chahl LA, Szolcsányi J, Lembeck F (eds) Antidromic vasodilatation and neurogenic inflammation. Akadémiai Kiadó, Budapest, pp 259-264

Chen RYZ, Li D-S, Guth PH (1992) Role of calcitonin gene-related peptide in capsaicin-induced gastric submucosal arteriolar dilation. Am J Physiol 262:H1350–H1355

Chéry-Croze S, Bosshard A, Martin H, Cuber JC, Charnay Y, Chayvialle JA (1989) Peptide immunocytochemistry in afferent neurons from lower gut in rats. Peptides 9:873–881

Cheung LY, Moody FG, Reese RS (1975) Effect of aspirin, bile salt and ethanol on canine gastric mucosal blood flow. Surgery 77:786–792

Chiba T, Yamaguchi A, Yamatani T, Nakamura A, Morishita T, Inui T, Fukase M, Noda T, Fujita T (1989) Calcitonin gene-related peptide receptor antagonist human CGRP-(8-37). Am J Physiol 256:E331–E335

Chiba T, Kinoshita Y, Nakamura A, Nakata Y (1991) Role of endogenous calcitonin gene-related peptide (CGRP) in gastric mucosal protection in the rat. Gastroenterology 100: A634

Chung K, Klein CM, Coggeshall RE (1990) The receptive part of the primary afferent axon is most vulnerable to systemic capsaicin in adult rats. Brain Res 511:222–226

Church MK, Lowman MA, Robinson C, Holgate ST, Benyon RC (1989) Interaction of neuropeptides with human mast cells. Int Arch Allergy Appl Immunol 88:70–78

Clarke GD, Davison JS (1978) Mucosal receptors in the gastric antrum and small intestine of the rat with afferent fibres in the cervical vagus. J Physiol (Lond) 284:55–67

Coderre TJ, Basbaum AI, Levine JD (1989) Neural control of vascular permeability: interactions between primary afferents, mast cells, and sympathetic efferents. J Neurophysiol 62:48–58

Coggeshall RE (1980) Law of separation of function of the spinal roots. Physiol Rev 60: 716–755

Colpaert FC, Donnerer J, Lembeck F (1983) Effects of capsaicin on inflammation and on the substance P content of nervous tissues in rats with adjuvant arthritis. Life Sci 32: 1827–1834

Constantine JW, Lebel WS, Woody HA (1991) Inhibition of tachykinin-induced hypotension in dogs by CP-96,345, a selective blocker of NK-1 receptors. Naunyn Schmiedebergs Arch Pharmacol 344: 471–477

Costa M, Furness JB (1984) Somatostatin is present in a subpopulation of noradrenergic nerve fibres supplying the intestine. Neuroscience 13:911–919

Costa M, Furness JB, Gibbins IL (1986) Chemical coding of enteric neurons. Prog Brain Res 68:217–239

Coutts AA, Jorizzo JL, Greaves MW, Burnstock G (1981) Mechanism of vasodilatation due to substance P in humans. Br J Dermatol 105:354–355

Couture R, Cuello AC (1984) Studies on the trigeminal antidromic vasodilatation and plasma extravasation in the rat. J Physiol (Lond) 346:273–285

Couture R, Kérouac R (1998) Plasma extravasation induced by mammalian tachykinins in rat skin: influence of anesthetic agents and an acetylcholine antagonist. Br J Pharmacol 91:265–273

Couture R, Cuello AC, Henry JL (1985) Trigeminal antidromic vasodilatation and plasma extravasation in the rat: effects of sensory, autonomic and motor denervation. Brain Res 346:108–114

Couture R, Lanéuville O, Guimond C, Drapeau G, Regoli D (1989) Characterization of the peripheral action of neurokinins and neurokinin receptor selective agonists on the rat cardiovascular system. Naunyn Schmiedebergs Arch Pharmacol 340:547–557

Crayton SC, Mitchell JH, Payne FC (1981) Reflex cardiovascular response during injection of capsaicin into skeletal muscle. Am J Physiol 240:H315–H319

Crossman DC, McEwan J, MacDermot J, MacIntyre I, Dollery CT (1987) Human calcitonin gene-related peptide activates adenylate cyclase and releases prostacyclin from human umbilical vein endothelium. Br J Pharmacol 92:695–702

Crossman DC, Dashwood MR, Brain SD, McEwan J, Pearson JD (1990) Action of calcitonin gene-related peptide upon bovine vascular endothelial and smooth muscle cells grown in isolation and co-culture. Br J Pharmacol 99:71–76

Cuello AC, Del Fiacco M, Paxinos G (1978) The central and peripheral ends of the substance P-containing sensory neurones in the rat trigeminal system. Brain Res 152: 499–509

Dalsgaard C-J (1988) The sensory system. In: Björklund A, Hökfelt T, Owman C (eds) The peripheral nervous system. Elsevier, Amsterdam, pp 599–636 (Handbook of chemical neuroanatomy, vol 6)

Dalsgaard C-J, Jonsson C-E, Hökfelt T, Cuello AC (1983) Localization of substance P-immunoreactive nerve fibers in the human digital skin. Experientia 39:1018–1020

Decker MW, Towle AC, Bissette G, Mueller RA, Lauder JM, Nemeroff CB (1985) Bombesin-like immunoreactivity in the central nervous system of capsaicin-treated rats: a radioimmunoassay and immunohistochemical study. Brain Res 342:1–8

Deguchi M, Niwa M, Shigematsu K, Fujii T, Namba K, Ozaki M (1989) Specific [^{125}I]Bolton-Hunter substance P binding sites in human and rat skin. Neurosci Lett 99: 287–292

Del Bianco E, Perretti F, Tramontana M, Manzini S, Geppetti P (1991) Calcitonin gene-related peptide in rat arterial and venous vessels: sensitivity to capsaicin, bradykinin and FMLP. Agents Actions 34: 376–380

Della NG, Papka RE, Furness JB, Costa M (1983) Vasoactive intestinal peptide-like immunoreactivity in nerves associated with the cardiovascular system of guinea pigs. Neuroscience 9:605–619

Dennis T, Fournier A, St Pierre S, Quirion R (1989) Structure-activity profile of calcitonin gene-related peptide in peripheral and brain tissues. Evidence for receptor multiplicity. J Pharmacol Exp Ther 251:718–725

Dennis T, Fournier A, Cadieux A, Pomerleau F, Jolicoeur FB, St Pierre S, Quirion R (1990) hCGRP$_{8-37}$, a calcitonin gene-related peptide antagonist revealing calcitonin gene-related peptide receptor heterogeneity in brain and periphery. J Pharmacol Exp Ther 254: 123–128

Devillier P, Regoli D, Asseraf A, Descours B, Marsac J, Renoux M (1986a) Histamine release and local responses of rat and human skin to substance P and other mammalian tachykinins. Pharmacology 32:340–347

Devillier P, Weill B, Renoux M, Menkes C, Pradelles P (1986b) Elevated levels of tachykinin-like immunoreactivity in joint fluids from patients with rheumatic inflammatory diseases. N Engl J Med 314:1323

Devor M, Papir-Kricheli D, Nachmias E, Rosenthal F, Gilon C, Chorev M, Selinger Z (1989) Substance P-induced cutaneous plasma extravasation in rats is mediated by NK-1 tachykinin receptors. Neurosci Lett 103:203–208

Dib B (1983) Dissociation between peripheral and central heat loss mechanisms induced by neonatal capsaicin. Behav Neurosci 97:822–829

Diez Guerra FJ, Zaidi M, Bevis P, MacIntyre I, Emson PC (1988) Evidence for release of calcitonin gene-related peptide and neurokinin A from sensory nerve endings in vivo. Neuroscience 25:839–846

DiPette DJ, Schwarzenberger K, Kerr N, Holland OB (1987) Systemic and regional hemodynamic effects of calcitonin gene-related peptide. Hypertension 9 [Suppl III]:III-142–III-146

Donnerer J, Lembeck F (1982) Analysis of the effects of intravenously injected capsaicin in the rat. Naunyn Schmiedebergs Arch Pharmacol 320:54–57

Donnerer J, Lembeck F (1983) Heat loss reaction through a peripheral site of action. Br J Pharmacol 79:719–723

Donnerer J. Stein C (1992) Evidence for an increase in the release of CGRP from sensory nerves during inflammation. Ann New York Acad Sci 657:505–506

Donnerer J, Barthó L, Holzer P, Lembeck F (1984) Intestinal peristalsis associated with release of immunoreactive substance P. Neuroscience 11:913–918

Donnerer J, Schuligoi R, Lembeck F (1989) Influence of capsaicin-induced denervation on neurogenic and humoral control of arterial pressure. Naunyn Schmiedebergs Arch Pharmacol 340:740–743

Donnerer J, Eglezos A, Helme RD (1990) Neuroendocrine and immune function in the capsaicin-treated rat: evidence for afferent neural modulation in vivo. In: Freier S (ed) The neuroendocrine-immune network. CRC Press, Boca Raton, pp 69–83

Donnerer J, Amann R, Lembeck F (1991) Neurogenic and non-neurogenic inflammation in the rat paw following chemical sympathectomy. Neuroscience 45:761–765

Donoso MV, Fournier A, St Pierre S, Huidobro-Toro P (1990) Pharmacological characterization of CGRP1 receptor subtype in the vascular system of the rat: studies with hCGRP fragments and analogs. Peptides 11:885–889

D'Orléans-Juste P, Dion S, Mizrahi J, Regoli D (1985) Effects of peptides and non-peptides on isolated arterial smooth muscles: role of endothelium. Eur J Pharmacol 114:9–21

D'Orleans-Juste P, Dion S, Drapeau G, Regoli D (1986) Different receptors are involved in the endothelium-mediated relaxation and the smooth muscle contraction of the rabbit pulmonary artery in response to substance P and related neurokinins. Eur J Pharmacol 125:37–44

D'Orléans-Juste P, Claing A, Télémaque S, Warner TD, Regoli D (1991) Neurokinins produce selective venoconstriction via NK-3 receptors in the rat mesenteric vascular bed. Eur J Pharmacol 204:329–334

Drapeau G, Rouissi N, Nantel F, Rhaleb N-E, Tousignant C, Regoli D (1990) Antagonists for the neurokinin NK-3 receptor evaluated in selective receptor systems. Regul Pept 31: 125–135

Duckles SP (1986) Effects of capsaicin on vascular smooth muscle. Naunyn Schmiedebergs Arch Pharmacol 333:59–64

Dunning BE, Taborsky GJ (1987) Calcitonin gene-related peptide: a potent and selective stimulator of gastrointestinal somatostatin secretion. Endocrinology 120:1774–1781

Ebertz JM, Hirshman CA, Kettelkamp NS, Uno H, Hanifin JM (1987) Substance P-induced histamine release in human cutaneous mast cells. J Invest Dermatol 88:682–685

Edvinsson L, Jansen I, Kingman TA, McCulloch J (1990) Cerebrovascular responses to capsaicin in vitro and in situ. Br J Pharmacol 100:312–318

Eglezos A, Giuliani S, Viti G, Maggi CA (1991) Direct evidence that capsaicin-induced plasma protein extravasation is mediated through tachykinin NK$_1$ receptors. Eur J Pharmacol 209:277–279

Ekblad E, Ekelund M, Graffner H, Håkanson R, Sundler F (1985) Peptide-containing nerve fibers in the stomach wall of rat and mouse. Gastroenterology 89:73–85

Eklund B, Jogestrand T, Pernow B (1977) Effect of substance P on resistance and capacitance vessels in the human forearm. In: von Euler US, Pernow P (eds) Substance P. Raven, New York, pp 275–285

Emonds-Alt X, Vilain P, Goulaouic P, Proietto V, Van Broeck D, Advenier C, Naline E, Neliat G, Le Fur G, Brelière JC (1992) A potent and selective non-peptide antagonist of the neurokinin A (NK$_2$) receptor. Life Sci 50: PL101–PL106

Endoh K, Leung FW (1990) Topical capsaicin protects the distal but not the proximal colon against acetic acid injury. Gastroenterology 98:A446

Erjavec F, Lembeck F, Florjanc-Irman T, Skofitsch G, Donnerer J, Saria A, Holzer P (1981) Release of histamine by substance P. Naunyn Schmiedebergs Arch Pharmacol 317:67–70

Esplugues JV, Whittle BJR (1990) Morphine potentiation of ethanol-induced gastric mucosal damage in the rat. Role of local sensory afferent neurons. Gastroenterology 98:82–89

Esplugues JV, Whittle BJR, Moncada S (1989) Local opioid-sensitive afferent sensory neurons in the modulation of gastric damage induced by Paf. Br J Pharmacol 97:579–585

Esplugues JV, Ramos EG, Gil L, Esplugues J (1990) Influence of capsaicin-sensitive afferent neurones on the acid secretory responses of the rat stomach in vivo. Br J Pharmacol 100:491–496

Evangelista S, Meli A (1989) Influence of capsaicin-sensitive fibres on experimentally-induced colitis in rats. J Pharm Pharmacol 41:574–575

Evangelista S, Maggi CA, Meli A (1986) Evidence for a role of adrenals in the capsaicin-sensitive "gastric defence mechanism" in rats. Proc Soc Exp Biol Med 182:568–569

Evangelista S, Lippe IT, Rovero P, Maggi CA, Meli A (1989) Tachykinins protect against ethanol-induced gastric lesions in rats. Peptides 10:79–81

Eysselein VE, Reinshagen M, Cominelli F, Sternini C, Davis W, Patel A, Nast CC, Bernstein D, Anderson K, Khan H, Snape WK (1991a) Calcitonin gene-related peptide and substance P decrease in the rabbit colon during colitis – a time study. Gastroenterology 101:1211–1219

Eysselein VE, Reinshagen M, Patel A, Davis W, Nast CC, Sternini C (1991b) CGRP in inflammatory bowel disease (IBD) and experimentally induced colitis. Regul Pept 34:88

Farber EM, Nickoloff BJ, Recht B, Fraki JE (1986) Stress, symmetry, and psoriasis: possible role of neuropeptides. J Am Acad Dermatol 14:305–311

Fearn HJ, Karady S, West GB (1965) The role of the nervous system in local inflammatory responses. J Pharm Pharmacol 17:761–765

Fehér E, Burnstock G, Varndell IM, Polak JM (1986) Calcitonin gene-related peptide-immunoreactive nerve fibres in the small intestine of the guinea pig: electron-microscopic immunocytochemistry. Cell Tissue Res 245:353–358

Ferrara A, Zinner MJ, Jaffe BM (1987) Intraluminal release of serotonin, substance P, and gastrin in the canine small intestine. Dig Dis Sci 32:289–294

Ferrell WR, Russell NJW (1986) Extravasation in the knee induced by antidromic stimulation of articular C fibre afferents of the anaesthetized cat. J Physiol (Lond) 379:407–416

Fewtrell CMS, Foreman JC, Jordan CC, Oehme P, Renner H, Stewart JM (1982) The effects of substance P on histamine and 5-hydroxytryptamine release in the rat. J Physiol (Lond) 330:393–411

Fiscus RR, Zhou H-L, Wang X, Han C, Ali S, Joyce SD, Murad F (1991) Calcitonin gene-related peptide (CGRP)-induced cyclic AMP, cyclic GMP and vasorelaxant responses in the rat thoracic aorta are antagonized by blockers of endothelium-derived relaxant factor (EDRF). Neuropeptides 20:133–143

Foreman JC, Jordan CC (1984) Neurogenic inflammation. Trends Pharmacol Sci 5:116–119

Foreman JC, Jordan CC, Oehme P, Renner H (1983) Structure-activity relationships for some substance P-related peptides that cause weal and flare reactions in human skin. J Physiol (Lond) 335:449–465

Forster ER, Dockray GJ (1991) The role of calcitonin gene-related peptide in gastric mucosal protection in the rat. Exp Physiol 76:623–626

Forster ER, Green T, Elliot M, Bremner A, Dockray GJ (1990) Gastric emptying in rats: role of afferent neurons and cholecystokinin. Am J Physiol 258:G552–G556

Franco-Cereceda A, Lundberg JM (1989) Post-occlusive reactive hyperaemia in the heart, skeletal muscle and skin of control and capsaicin-pre-treated pigs. Acta Physiol Scand 137:271–277

Franco-Cereceda A, Gennari C, Nami R, Agnusdei D, Pernow J, Lundberg JM, Fischer JA (1987a) Cardiovascular effects of calcitonin gene-related peptides I and II in man. Circ Res 60:393–397

Franco-Cereceda A, Henke H, Lundberg JM, Petermann JB, Hökfelt T, Fischer JA (1987b) Calcitonin gene-related peptide (CGRP) in capsaicin-sensitive substance P-immunoreactive sensory neurons in animals and man: distribution and release by capsaicin. Peptides 8:399–410

Frossard N, Rhoden KJ, Barnes PJ (1989) Influence of epithelium on guinea pig airway responses to tachykinins: role of endopeptidase and cyclooxygenase. J Pharmacol Exp Ther 248:292–298

Fujimori A, Saito A, Kimura S, Watanabe T, Uchiyama Y, Kawasaki H, Goto K (1989) Neurogenic vasodilation and release of calcitonin gene-related peptide (CGRP) from perivascular nerves in the rat mesenteric artery. Biochem Biophys Res Commun 165:1391–1398

Fujimori A, Saito A, Kimura S, Goto K (1990) Release of calcitonin gene-related peptide (CGRP) from capsaicin-sensitive vasodilator nerves in the rat mesenteric artery. Neurosci Lett 112:173–178

Fujita S, Shimizu T, Izumi K, Fukuda T, Sameshima M, Ohba N (1984) Capsaicin-induced neuroparalytic keratitis-like corneal changes in the mouse. Exp Eye Res 38:165–175

Fuller RW, Conradson T-B, Dixon CMS, Crossman DC, Barnes PJ (1987) Sensory neuro-peptide effects in human skin. Br J Pharmacol 92:781–788

Furchgott RF (1984) The role of endothelium in the responses of vascular smooth muscle to drugs. Annu Rev Pharmacol Toxicol 24:175–197

Furness JB, Papka RE, Della NG, Costa M, Eskay RL (1982) Substance P-like immunore-activity in nerves associated with the vascular system of guinea-pigs. Neuroscience 7: 447–459

Galligan JJ, Costa M, Furness JB (1988) Changes in surviving nerve fibers associated with submucosal arteries following extrinsic denervation of the small intestine. Cell Tissue Res 253:647–656

Galligan JJ, Jiang M-M, Shen K-Z, Suprenant A (1990) Substance P mediates neurogenic vasodilatation in extrinsically denervated guinea-pig submucosal arterioles. J Physiol (Lond) 420:267–280

Gamse R, Saria A (1985a) The spinal cord contains multiple factors causing plasma protein extravasation in the skin. Eur J Pharmacol 113:363–371

Gamse R, Saria A (1985b) Potentiation of tachykinin-induced plasma protein extravasation by calcitonin gene-related peptide. Eur J Pharmacol 114:61–66

Gamse R, Saria A (1987) Antidromic vasodilatation in the rat hindpaw measured by laser Doppler flowmetry: pharmacological modulation. J Auton Nerv Syst 19:105–111

Gamse R, Holzer P, Lembeck F (1980) Decrease of substance P in primary sensory neuro-nes and impairment of neurogenic plasma extravasation by capsaicin. Br J Pharmacol 68: 207–213

Gamse R, Leeman SE, Holzer P, Lembeck F (1981) Differential effects of capsaicin on the content of somatostatin, substance P, and neurotensin in the nervous system of the rat. Naunyn Schmiedebergs Arch Pharmacol 317:140–148

Gao GC, Dashwood MR, Wei ET (1991) Corticotropin-releasing factor inhibition of sub-stance P-induced vascular leakage in rats: possible sites of action. Peptides 12:639–644

Garcia Leme J, Hamamura L (1974) Formation of a factor increasing vascular permeability during electrical stimulation of the saphenous nerve in rats. Br J Pharmacol 51:383–389

Gardiner SM, Compton AM, Bennett T (1989) Regional hemodynamic effects of calcitonin gene-related peptide. Am J Physiol 256:R332–R338

Gardiner SM, Compton AM, Kemp PA, Bennett T, Bose C, Foulkes C, Hughes B (1990) Antagonistic effect of human alpha-CGRP[8-37] on the in vivo regional haemodynamic actions of human alpha-CGRP. Biochem Biophys Res Commun 171:938–943

Garret C, Carruette A, Fardin V, Moussaoui S, Peyronel J-F, Blanchard J-C, Laduron PM (1991) Pharmacological properties of a potent and seletive nonpeptide substance P antag-onist. Proc Natl Acad Sci USA 88:10208–10212

Gates TS, Zimmerman RP, Mantyh CR, Vigna R, Maggio JE, Welton ML. Passaro EP, Mantyh PW (1988) Substance P and substance K receptor binding sites in the human gastrointestinal tract: localization by autoradiography. Peptides 9:1207–1219

Gates TS, Zimmerman RP, Mantyh CR, Vigna SR, Mantyh PW (1989) Calcitonin gene-related peptide-α receptor binding sites in the gastrointestinal tract. Neuroscience 31: 757–770

Gazelius B, Brodin E, Olgart L, Panopoulos P (1981) Evidence that substance P is a me-diator of antidromic vasodilatation using somatostatin as a release inhibitor. Acta Physiol Scand 113:155–159

Geppetti P, Frilli S, Renzi D, Santicioli P, Maggi CA, Theodorsson E, Fanciullacci M (1988) Distribution of calcitonin gene-related peptide-like immunoreactivity in various rat tissues: correlation with substance P and other tachykinins and sensitivity to capsaicin. Regul Pept 23:289–298

Geppetti P, Santicioli P, Rubini I, Spillantini MG, Maggi CA, Sicuteri F (1989) Thiorphan increases capsaicin-evoked release of substance P from slices of dorsal spinal cord of guinea pig. Neurosci Lett 103:69–73

Geppetti P, Tramontana M, Evangelista S, Renzi D, Maggi CA, Fusco BM, Del Bianco E (1991) Differential effect on neuropeptide release of different concentrations of hydrogen ions on afferent and intrinsic neurons of the rat stomach. Gastroenterology 101: 1505–1511

Gibbins IL, Furnes JB, Costa M, MacIntyre I, Hillyard CJ, Girgis S (1985) Co-localization of calcitonin gene-related peptide-like immonoreactivity with substance P in cutaneous, vascular and visceral sensory neurons of guinea pigs. Neurosci Lett 57:125–130

Gibbins IL, Furness JB, Costa M (1987a) Pathway-specific patterns of the co-existence of substance P, calcitonin gene-related peptide, cholecystokinin and dynorphin in neurons of the dorsal root ganglia of the guinea-pig. Cell Tissue Res 248:417–437

Gibbins IL, Wattchow D, Coventry B (1987b) Two immunohistochemically identified populations of calcitonin gene-related peptide (CGRP)-immunoreactive axons in human skin. Brain Res 414:143–148

Gibbins IL, Morris JL, Furness JB, Costa M (1988) Innervation of systemic blood vessels. In: Burnstock G, Griffith SG (eds) Nonadrenergic innervation of blood vessels: II. Regional innervation. CRC Press, Boca Raton, pp 1–36

Giuliani S, Santicioli P, Tramontana M, Geppetti P, Maggi CA (1991) Peptide N-formyl-methionyl-leucyl-phenylalanine (FMLP) activates capsaicin-sensitive afferent nerves in guinea-pig atria and urinary bladder. Br J Pharmacol 102:730–734

Goehler LE, Sternini C, Brecha NC (1988) Calcitonin gene-related peptide immunoreactivity in the biliary pathway and liver of the guinea-pig: distribution and colocalization with substance P. Cell Tissue Res 253:145–150

Grace GC, Dusting GJ, Kemp BE, Martin TJ (1987) Endothelium and the vasodilator action of rat calcitonin gene-related peptide (CGRP). Br J Pharmacol 91:729–733

Gray JL, Brunnet NW, Mulvihill SJ, Debas HT (1989) Capsaicin stimulates release of calcitonin gene-related peptide (CGRP) from the isolated perfused rat stomach. Gastroenterology 96:A181

Green T, Dockray GJ (1988) Characterization of the peptidergic afferent innervation of the stomach in the rat, mouse, and guinea-pig. Neuroscience 25:181–193

Greenberg B, Rhoden K, Barnes P (1987) Calcitonin gene-related peptide (CGRP) is a potent non-endothelium-dependent inhibitor of coronary vasomotor tone. Br J Pharmacol 92:789–794

Greenway CV (1983) Role of splanchnic venous system in overall cardiovascular homeostasis. Fed Proc 42:1678–1684

Grönstad KO, Ahlman H, Zinner MJ, Jaffe BM (1983) The effect of vagal nerve stimulation on feline portal venous levels of substance P. In: Skrabanek P, Powell D (eds) Substance P Dublin 1983. Boole, Dublin, pp 153–154

Grönstad KO, Nilsson O, Dahlström A, Price B, Zinner MJ, Jaffe BM, Ahlman H (1985) The effects of vagal nerve stimulation on endoluminal release of serotonin and substance P into the feline small intestine. Scand J Gastroenterol 20:163–169

Grönstad KO, Dahlström A, Jaffe BM, Zinner MJ, Ahlman H (1986) Studies on the mucosal hyperaemia of the feline small intestine observed at endoluminal perfusion with substance P. Acta Physiol Scand 128:97–108

Guard S, Watson SP (1991) Tachykinin receptor types: classification and membrane signalling mechanisms. Neurochem Int 18:149–165

Guth PH, Leung FW, Kauffman GL (1989) Physiology of the gastric circulation. In: Schultz SG (ed) The gastrointestinal system. American Physiological Society, Bethesda, pp 1371–1404 (Handbook of physiology, section 6, vol I, part 2)

Haegerstrand A, Dalsgaard C-J, Jonzon B, Larsson O, Nilsson J (1990) Calcitonin gene-related peptide stimulates proliferation of human endothelial cells. Proc Natl Acad Sci USA 87:3299–3303

Hagan RM, Ireland SJ, Bailey F, McBride C, Jordan CC, Ward P (1991) A spirolactam conformationally-restrained analogue of physalaemin which is a peptidase-resistant, selective neurokinin NK_1 receptor antagonist. Br J Pharmacol 102:168P

Hagen NA, Stevens JC, Michet CJ (1990) Trigeminal sensory neuropathy associated with connective tissue diseases. Neurology 40:891–896

Hägermark O, Hökfelt T, Pernow B (1978) Flare and itch induced by substance P in human skin. J Invest Dermatol 71:233–235

Hajós M, Obál F, Jancsó G, Obál F (1983) The capsaicin sensitivity of the preoptic region is preserved in adult rats pretreated as neonates, but lost in rats pretreated as adults. Naunyn Schmiedebergs Arch Pharmacol 324:219–222

Håkanson R, Sundler F (1985) Tachykinin antagonists. Elsevier, Amsterdam

Håkanson R, Leander S, Andersson RGG, Hörig J (1983) Substance P antagonists release histamine from peritoneal mast cells. Acta Physiol Scand 117:319–320

Hall JM, Fox AJ, Morton IKM (1990) Peptidase activity as a determinant of agonist potencies in some smooth muscle preparations. Eur J Pharmacol 176:127–134

Hall ME, Miley F, Stewart JM (1989) The role of enzymatic processing in the biological actions of substance P. Peptides 10:895–901

Hallberg D, Pernow B (1975) Effect of substance P on various vascular beds in the dog. Acta Physiol Scand 93:277–285

Han S-P, Naes L, Westfall TC (1990a) Inhibition of periarterial nerve stimulation-induced vasodilation of the mesenteric arterial bed by CGRP (8-37) and CGRP receptor desensitization. Biochem Biophys Res Commun 168:786–791

Han S-P, Naes L, Westfall TC (1990b) Calcitonin gene-related peptide is the endogenous mediator of nonadrenergic noncholinergic vasodilation in rat mesentery. J Pharmacol Exp Ther 255:423–428

Hartschuh W, Weihe E, Reinecke M (1983) Peptidergic (neurotensin, VIP, substance P) nerve fibres in the skin. Immunohistochemical evidence for an involvement of neuropeptides in nociception, pruritus and inflammation. Br J Dermatol 109 [Suppl 25]:14–17

Hartung H-P, Wolters K, Toyka KV (1986) Substance P: binding properties and studies on cellular responses in guinea-pig macrophages. J Immunol 136:3856–3863

Helke CJ, Niederer AJ (1990) Studies on the coexistence of substance P with other putative transmitters in the nodose and petrosal ganglia. Synapse 5:144–151

Hellström PM, Söder O, Theodorsson E (1991) Occurrence, release, and effects of multiple tachykinins in cat colonic tissue and nerves. Gastroenterology 100:431–440

Helme RD, Andrews PV (1985) The effect of nerve lesions on the inflammatory response to injury. J Neurosci Res 13:453–459

Helme RD, McKernan S (1984) Flare responses in man following topical application of capsaicin. In: Chahl LA, Szolcsányi J, Lembeck F (eds) Antidromic vasodilatation and neurogenic inflammation. Akadémiai Kiadó, Budapest, pp 303–315

Helme RD, McKernan S (1985) Neurogenic flare responses following topical application of capsaicin in humans. Ann Neurol 18:505–509

Helme RD, Koschorke GM, Zimmermann M (1986) Immunoreactive substance P release from skin nerves in the rat by noxious thermal stimulation. Neurosci Lett 63:295–299

Helme RD, Eglezos A, Hosking CS (1987) Substance P induces chemotaxis of neutrophils in normal and capsaicin-treated rats. Immunol Cell Biol 65:267–269

Helton WS, Mulholland MM, Bunnett NW, Debas HT (1989) Inhibition of gastric and pancreatic secretion in dogs by CGRP: role of somatostatin. Am J Physiol 256: G715–G720

Henriksen JH, Bülow JB, Schaffalitzky de Muckadell O, Fahrenkrug J (1986) Do substance P and vasoactive intestinal polypeptide (VIP) play a role in the acute occlusive or chronic ischaemic vasodilation in man? Clin Physiol 6:163–170

Heyer G, Hornstein OP, Handwerker HO (1991) Reactions to intradermally injected substance P and topically applied mustard oil in atopic dermatitis patients. Acta Derm Venereol (Stock) 71:291–295

Hinsey JC, Gasser HS (1930) The component of the dorsal root mediating vasodilatation and the Sherrington contracture. Am J Physiol 92:679–689

Hirata Y, Takagi Y, Takata S, Fukuda Y, Yoshima H, Fujita T (1988) Calcitonin gene-related peptide receptor in cultured vascular smooth muscle and endothelial cells. Biochem Biophys Res Commun 151:1113–1121

Hökfelt T, Kellerth J-O, Nilsson G, Pernow B (1975) Experimental immunohistochemical studies on the localization and distribution of substance P in cat primary sensory neurons. Brain Res 100:235–252

Holliger C, Radzyner M, Knoblauch M (1983) Effects of glucagon, vasoactive intestinal peptide, and vasopressin on villous microcirculation and superior mesenteric artery blood flow of the rat. Gastroenterology 85:1036–1043

Holmdahl G, Håkanson R, Leander S, Rosell S, Folkers K, Sundler F (1981) A substance P antagonist, [D-Pro^2,D-Trp^{7,9}]SP, inhibits inflammatory responses in the rabbit eye. Science 214:1029–1031

Holton P, Perry WLM (1951) On the transmitter responsible for antidromic vasodilatation in the rabbit's ear. J Physiol (Lond) 114:240–251

Holzer P (1984) Characterization of the stimulus-induced release of immunoreactive substance P from the myenteric plexus of the guinea-pig small intestine. Brain Res 297:127–136

Holzer P (1988) Local effector functions of capsaicin-sensitive sensory nerve endings: involvement of tachykinins, calcitonin gene-related peptide and other neuropeptides. Neuroscience 24:739–768

Holzer P (1991) Capsaicin: cellular targets, mechanisms of action, and selectivity for thin sensory neurons. Pharmacol Rev 43:143–201

Holzer P, Guth PH (1991) Neuropeptide control of rat gastric mucosal blood flow. Increase by calcitonin gene-related peptide and vasoactive intestinal polypeptide, but not substance P and neurokinin A. Circ Res 68:100–105

Holzer P, Lippe IT (1988) Stimulation of afferent nerve endings by intragastric capsaicin protects against ethanol-induced damage of gastric mucosa. Neuroscience 27:981–987

Holzer P, Lippe IT (1992) Gastric mucosal hyperemia due to acid backdiffusion depends on splanchnic nerve activity. Am J Physiol 262:G505–G509

Holzer P, Sametz W (1986) Gastric mucosal protection against ulcerogenic factors in the rat mediated by capsaicin-sensitive afferent neurons. Gastroenterology 91:975–981

Holzer P, Gamse R, Lembeck F (1980) Distribution of substance P in the rat gastrointestinal tract – lack of effect of capsaicin pretreatment. Eur J Pharmacol 61:303–307

Holzer P, Bucsics A, Lembeck F (1982) Distribution of capsaicin-sensitive nerve fibres containing immunoreactive substance P in cutaneous and visceral tissues of the rat. Neurosci Lett 31:253–257

Holzer P, Pabst MA, Lippe IT (1989) Intragastric capsaicin protects against aspirin-induced lesion formation and bleeding in the rat gastric mucosa. Gastroenterology 96:1425–1433

Holzer P, Pabst MA, Lippe IT, Peskar BM, Peskar BA, Livingston EH, Guth PH (1990a) Afferent nerve-mediated protection against deep mucosal damage in the rat stomach. Gastroenterology 98:838–848

Holzer P, Peskar BM, Peskar BA, Amann R (1990b) Release of calcitonin gene-related peptide induced by capsaicin in the vascularly perfused rat stomach. Neurosci Lett 108:195–200

Holzer P, Livingston EH, Guth PH (1991a) Sensory neurons signal for an increase in rat gastric mucosal blood flow in the face of pending acid injury. Gastroenterology 101:416–423

Holzer P, Livingston EH, Saria A, Guth PH (1991b) Sensory neurons mediate protective vasodilatation in rat gastric mucosa. Am J Physiol 260:G363–G370

Holzer P, Lippe IT, Amann R (1992) Participation of capsaicin-sensitive afferent neurons in gastric motor inhibition caused by laparotomy and intraperitoneal acid. Neuroscience 48:715–722

Holzer-Petsche U, Schimek E, Amann R, Lembeck F (1985) In vivo and in vitro actions of mammalian tachykinins. Naunyn Schmiedebergs Arch Pharmacol 330:130–135

Hornyak ME, Naver HK, Rydenhag B, Wallin BG (1990) Sympathetic activity influences the vascular axon reflex in the skin. Acta Physiol Scand 139:77–84

Hottenstein OD, Kreulen DL (1987) Comparison of the frequency dependence of venous and arterial responses to sympathetic nerve stimulation in guinea-pigs. J Physiol (Lond) 384:153–167

Hottenstein OD, Remak G, Jacobson ED (1990) Post-nerve stimulation hyperemia in rat intestinal circulation involves capsaicin-sensitive nerves. Gastroenterology 98:A175

Hottenstein OD, Pawlik WW, Remak G, Jacobson ED (1991) Capsaicin-sensitive nerves modulate resting blood flow and vascular tone in rat gut. Naunyn Schmiedebergs Arch Pharmacol 343:179–184

Hottenstein OD, Pawlik WW, Remak G, Jacobson ED (1992) Capsaicin-sensitive nerves modulate reactive hyperemia in rat gut. Proc Soc Exp Biol Med 199:311–320

Hua X-Y, Lundberg JM, Theodorsson-Norheim E, Brodin E (1984) Comparison of cardiovascular and bronchoconstrictor effects of substance P, substance K and other tachykinins. Naunyn Schmiedebergs Arch Pharmacol 328:196–201

Hua X-Y, Theodorsson-Norheim E, Brodin E, Lundberg JM, Hökfelt T (1985) Multiple tachykinins (neurokinin A, neuropeptide K and substance P) in capsaicin-sensitive sensory neurons in the guinea-pig. Regul Pept 13:1–19

Huang M, Rorstad OP (1987) VIP receptors in mesenteric and coronary arteries: a radioligand study. Peptides 8:477–485

Hughes SR, Brain SD (1991) A calcitonin gene-related peptide (CGRP) antagonist (CGRP$_{8-37}$) inhibits microvascular responses induced by CGRP and capsaicin in skin. Br J Pharmacol 104:738–742

Hughes SR, Williams TJ, Brain SD (1990) Evidence that endogenous nitric oxide modulates oedema formation induced by substance P. Eur J Pharmacol 191:481–484

Hutcheson IR, Whittle BJR, Boughton-Smith NK (1990) Role of nitric oxide in maintaining vascular integrity in endotoxin-induced acute intestinal damage in the rat. Br J Pharmacol 101:815–820

Inui T, Chiba T, Okimura Y, Morishita T, Nakamura A, Yamaguchi A, Yamatani T, Kadowaki S, Chihara K, Fujita T (1989) Presence and release of calcitonin gene-related peptide in rat stomach. Life Sci 45:1199–1206

Ishida-Yamamoto A, Senba A, Tohyama M (1989) Distribution and fine structure of calcitonin gene-related peptide-like immunoreactive nerve fibers in the rat skin. Brain Res 491:93–101

Ito S, Ohga A, Ohta T (1988) Gastric vasodilatation and vasoactive intestinal peptide output in response to vagal stimulation in the dog. J Physiol (Lond) 404:669–682

Iwamoto I, Nadel JA (1989) Tachykinin receptor subtype that mediates the increase in vascular permeability in guinea pig skin. Life Sci 44:1089–1095

Iwamoto I, Kimura A, Tanaka M, Tomioka H, Yoshida S (1990) Skin reactivity to substance P, not to neurokinin A, is increased in allergic asthmatics. Int Arch Allergy Appl Immunol 93:120–125

Jacques L, Couture R, Drapeau G, Regoli D (1989) Capillary permeability induced by intravenous neurokinins: receptor characterization and mechanisms of action. Naunyn Schmiedebergs Arch Pharmacol 340:170–179

Jancsó G (1984) Sensory nerves as modulators of inflammatory reactions. In: Chahl LA, Szolcsányi J, Lembeck F (eds) Antidromic vasodilatation and neurogenic inflammation. Akadémiai Kiadó, Budapest, pp 207–222

Jancsó G, Király E, Jancsó-Gábor A (1977) Pharmacologically induced selective degeneration of chemosensitive primary sensory neurones. Nature 270:741–743

Jancsó G, Király E, Jancsó-Gábor A (1980) Chemosensitive pain fibres and inflammation. Int J Tissue React 2:57–66

Jancsó G, Hökfelt T, Lundberg JM, Király E, Halász N, Nilsson G, Terenius L, Rehfeld J, Steinbusch H, Verhofstad AER, Said S, Brown M (1981) Immunohistochemical studies on the effect of capsaicin on spinal and medullary peptide and monoamine neurons using antisera to substance P, gastrin/CCK, somatostatin, VIP, enkephalin, neurotensin and 5-hydroxytryptamine. J Neurocytol 10:963–980

Jancsó G, Husz S, Simon N (1983) Impairment of axon reflex vasodilatation after herpes zoster. Clin Exp Dermatol 8:27–31

Jancsó G, Obál F, Tóth-Kása I, Katona M, Husz S (1985) The modulation of cutaneous inflammatory reactions by peptide-containing sensory nerves. Int J Tissue React 7:449–457

Jancsó G, Király E, Such G, Joó F, Nagy A (1987) Neurotoxic effect of capsaicin in mammals. Acta Physiol Hung 69:295–313

Jancsó N, Jancsó-Gábor A, Szolcsányi J (1967) Direct evidence for neurogenic inflammation and its prevention by denervation and by pretreatment with capsaicin. Br J Pharmacol 31:138–151

Jancsó N, Jancsó-Gábor A, Szolcsányi J (1968) The role of sensory nerve endings in neurogenic inflammation induced in human skin and in the eye and paw of the rat. Br J Pharmacol 33:32–41

Jänig W, Lisney SJW (1989) Small diameter myelinated afferents produce vasodilatation but not plasma extravasation in rat skin. J Physiol (Lond) 415:477–486

Jansen G, Lundeberg T, Kjartansson J, Samuelson UE (1989) Acupuncture and sensory neuropeptides increase cutaneous blood flow in rats. Neurosci Lett 97:305–309

Jessell TM, Iversen LL, Cuello AC (1978) Capsaicin-induced depletion of substance P from primary sensory neurones. Brain Res 152:183–188

Jin J-G, Nakayama S (1990) Bile salt potentiates the action of capsaicin on sensory neurones of guinea-pig ileum. Neurosci Lett 109:88–91

Jodal M, Lundgren O (1989) Neurohormonal control of gastrointestinal blood flow. In: Schultz SG (ed) The gastrointestinal system. American Physiological Society, Bethesda, pp 1667–1707 (Handbook of Physiology, section 6, vol I, part 2)

Johnson AR, Erdös EG (1973) Release of histamine from mast cells by vasoactive peptides. Proc Soc Exp Biol Med 142:1252–1256

Jonsson C-E, Brodin, E, Dalsgaard C-J, Haegerstrand A (1986) Release of substance-P-like immunoreactivity in dog paw lymph after scalding injury. Acta Physiol Scand 126:21–24

Jorizzo J, Coutts AA, Eady RAJ, Greaves MW (1983) Vascular responses of human skin to injection of substance P and mechanism of action. Eur J Pharmacol 87:67–76

Ju G, Hökfelt T, Brodin E, Fahrenkrug J, Fischer JA, Frey P, Elde RP, Brown JC(1987) Primary sensory neurons of the rat showing calcitonin gene-related peptide immunoreactivity and their relation to substance P-somatostatin-, galanin-, vasoactive intestinal polypeptide- and cholecystokinin-immunoreactive ganglion cells. Cell Tissue Res 247: 417–431

Juan H, Lembeck F (1974) Action of peptides and other algesic agents on paravascular pain receptors of the isolated perfused rabbit ear. Naunyn Schmiedebergs Arch Pharmacol 283:151–164

Kaada B, Olsen E, Eielsen O (1984) In search of mediators of skin vasodilation induced by transcutaneous nerve stimulation: III. Increase in plasma VIP in normal subjects and in Raynaud's disease. Gen Pharmacol 15:107–113

Kachelhoffer J, Eloy MR, Pousse A, Hohmatter D, Grenier JF (1974) Mesenteric vasomotor effects of vasoactive intestinal polypeptide. Study on perfused isolated canine jejunal loops. Pflügers Arch 352:37–46

Kakudo K, Hasegawa H, Komatsu N, Nakamura A, Itoh Y, Watanabe K (1988) Immuno-electron microscopic study oc calcitonin gene-related peptide (CGRP) in axis cylinders of the vagus nerve. CGRP is present in both myelinated and unmyelinated fibers. Brain Res 440:153–158

Karmeli F, Eliakim R, Okon E, Rachmilewitz D (1991) Gastric mucosal damage by ethanol is mediated by substance P and prevented by ketotifen, a mast cell stabilizer. Gastroenterology 100:1206–1216

Kashiba H, Senba E, Ueda Y, Tohyama M (1990) Relative sparing of calcitonin gene-related peptide-containing primary sensory neurons following neonatal capsaicin treatment in the rats. Peptides 11:491–496

Kashiba H, Senba E, Ueda Y, Tohyama M (1991) Cell size and cell type analysis of calcitonin gene-related peptide-containing cutaneous and splanchnic sensory neurons in the rat. Peptides 12:101–106

Kawasaki H, Takasaki K, Saito A, Goto K (1988) Calcitonin gene-related peptide acts as a novel vasodilator neurotransmitter in mesenteric resistance vessels of the rat. Nature 335: 164–167

Kawasaki H, Nuki C, Saito A, Takasaki K (1990a) Role of calcitonin gene-related peptide-containing nerves in the vascular adrenergic neurotransmission. J Pharmacol Exp Ther 252:403–409

Kawasaki H, Nuki C, Saito A, Takasaki K (1990b) Adrenergic modulation of calcitonin gene-related peptide (CGRP)-containing nerve-mediated vasodilation in the rat mesenteric resistance vessel. Brain Res 506:287–290

Kawasaki H, Saito A, Takasaki K (1990c) Age-related decrease of calcitonin gene-related peptide-containing vasodilator innervation in the mesenteric resistance vessel of the spontaneously hypertensive rat. Circ Res 67:733–743

Kawasaki H, Saito A, Takasaki K (1990d) Changes in calcitonin gene-related peptide (CGRP)-containing vasodilator nerve activity in hypertension. Brain Res 518:303–307

Kawasaki H, Nuki C, Saito A, Takasaki K (1991a) NPY modulates neurotransmission of CGRP-containing vasodilator nerves in rat mesenteric arteries. Am J Physiol 261: H683–H690

Kawasaki H, Saito A, Goto K, Takasaki K (1991b) Age-related changes in calcitonin gene-related peptide (CGRP)-mediated neurogenic vasodilation of the mesenteric resistance vessel in SHR. Clin Exp Hypertens [A] 13:745–754

Keast JR, Furness JB, Costa M (1985) Distribution of certain peptide-containing nerve fibres and endocrine cells in the gastrointestinal mucosa in five mammalian species. J Comp Neurol 236:403–422

Keen P, Harmar AJ, Spears F, Winter E (1982) Biosynthesis, axonal transport and turnover of neuronal substance P. In: Porter R, O'Connor M (eds) Substance P in the nervous system. Pitman, London, pp 145–160

Kenins P (1981) Identification of the unmyelinated sensory nerves which evoke plasma extravasation in response to antidromic stimulation. Neurosci Lett 25:137–141

Kenins P, Hurley JV, Bell C (1984) The role of substance P in the axon reflex in the rat. Br J Dermatol 111:551–559

Khalil Z, Helme RD (1989a) Involvement of capsaicin-sensitive afferent nerve fibres in serotonin-induced plasma extravasation and vasodilatation in rat skin. Neurosci Lett 104: 105–109

Khalil Z, Helme RD (1989b) Sequence of events in substance P-mediated plasma extravasation in rat skin. Brain Res 500:256–262

Khalil Z, Andrews PV, Helme RD (1988) VIP modulates substance P-induced plasma extravasation in vivo. Eur J Pharmacol 151:281–288

Khayutin VM, Sonina RS, Frolenkov GI, Zizin IM (1991) Antidromic vasodilation in frog: identification of the nerve fiber types involved. Pflügers Arch 419:508–513

Kiang JG, Wei ET (1985) CRF: an inhibitor of neurogenic plasma extravasation produced by saphenous nerve stimulation. Eur J Pharmacol 114:111–112

Kiernan JA (1971) Degranulation of mast cells following antidromic stimulation of cutaneous nerves. J Anat 111:349–350

Kiernan JA (1972) The involvement of mast cells in vasodilatation due to axon reflexes in injured skin. Q J Exp Physiol 57:311–317

Kiernan JA (1975) A pharmacological and histological investigation of the involvement of mast cells in cutaneous axon reflex vasodilatation. Q J Exp Physiol 60:123–130

Kiernan JA (1976) Evidence for the involvement of vasoactive constituents of mast cells in axon reflex vasodilatation in the skin of the rat. Arch Dermatol Res 255:1–8

Kiernan JA (1984) Role of mast cells in vascular responses due to axon reflexes in injured skin. In: Chahl LA, Szolcsányi J, Lembeck L (eds) Antidromic vasodilatation and neurogenic inflammation. Akadémiai Kiadó, Budapest, pp 223–243

Kirchgessner AL, Dodd J, Gershon MD (1988) Markers shared between dorsal root and enteric ganglia. J Comp Neurol 276:607–621

Kjartansson J, Dalsgaard C-J, Jonsson C-E (1987) Decreased survival of experimental critical flaps in rats after sensory denervation with capsaicin. Plast Reconstr Surg 79: 218–221

Kjartansson J, Lundberg T, Samuelson UE, Dalsgaard C-J, Hedén P (1988) Calcitonin gene-related peptide (CGRP) and transcutaneous electrical nerve stimulation (TENS) increase cutaneous blood flow in a musculocutaneous flap in the rat. Acta Physiol Scand 134:89–94

Knyazev GG, Knyazeva GB, Nikiforov AF (1990) Neuroparalytic keratitis and capsaicin. Acta Physiol Hung 75:29–34

Knyazev GG, Knyazeva GB, Tolochko ZS (1991) Trophic functions of primary sensory neurons: are they really local? Neuroscience 42:555–560

Kocher L, Anton F, Reeh PW, Handwerker HO (1987) The effect of carrageenan-induced inflammation on the sensitivity of unmyelinated skin nociceptors in the rat. Pain 29: 363–373

Koltzenburg M, Lewin G, McMahon S (1990) Increase of blood flow in skin and spinal cord following activation of small diameter primary afferents. Brain Res 509:145–149

Kolve E, Taché Y (1989) Intracisternal α-CGRP prevents gastric ulcer formation in the rat. Gastroenterology 96:A266

Kondo H, Yamamoto M (1988) The ontogeny and fine structure of calcitonin gene-related peptide (CGRP)-immunoreactive nerve fibers in the celiac ganglion of rats. Arch Histol Cytol 51:91–98

Kondo H, Yui R (1981) An electron microscopic study on substance P-like immunoreactive nerve fibers in the celiac ganglion of guinea pigs. Brain Res 222:134–137

Konishi S, Tsunoo A, Yanaihara N, Otsuka M (1980) Peptigergic excitatory and inhibitory synapses in mammalian sympathetic ganglia: roles of substance P and enkephalin. Biomed Res 1:528–536

Kowalski ML, Kaliner MA (1988) Neurogenic inflammation, vascular permeability, and mast cells. J Immunol 140:3905–3911

Kowalski ML, Sliwinska-Kowalska M, Kaliner MA (1990) Neurogenic inflammation, vascular permeability, and mast cells. 2. Additional evidence indicating that mast cells are not involved in neurogenic inflammation. J Immunol 145:1214–1221

Kreulen DL (1986) Activation of mesenteric arteries and veins by preganglionic and postganglionic nerves in the guinea-pig. Am J Physiol 251:H1267–H1275

Kubota M, Moseley JM, Butera L, Dusting GJ, MacDonald PS, Martin TJ (1985) Calcitonin gene-related peptide stimulates cyclic AMP formation in rat aortic smooth muscle cells. Biochem Biophys Res Commun 132:88–94

Kuo DC, Kawatani M, de Groat WC (1985) Vasoactive intestinal polypeptide identified in the thoracic dorsal root ganglia of the cat. Brain Res 330:178–182

Kurozawa Y, Nasu Y, Nose T (1991) Response of capsaicin pretreated skin blood vessels to exercise. Acta Physiol Scand 141:181–184

Kwok YN, McIntosh CHS (1990) Release of substance P-like immunoreactivity from the vascularly perfused rat stomach. Eur J Pharmacol 180:201–207

Lacroix JS, Buvelot JM, Polla BS, Lundberg JM (1991) Improvement of symptoms of non-allergic chronic rhinitis by local treatment with capsaicin. Clin Exp Allergy 21:595–600

Lang E, Novak A, Reeh PW, Handwerker HO (1990) Chemosensitivity of fine afferents from rat skin in vitro. J Neurophysiol 63:887–901

Lee C-M, Sandberg BEB, Hanley MR, Iversen LL (1981) Purification and characterization of a membrane-bound substance P-degrading enzyme from human brain. Eur J Biochem 114:315–327

Lee Y, Takami K, Kawai Y, Girgis S, Hillyard CJ, MacIntyre I, Emson PC, Tohyama M (1985) Distribution of calcitonin gene-related peptide in the rat peripheral nervous system with reference to its coexistence with substance P. Neuroscience 15:1227–1237

Lee Y, Hayashi N, Hillyard CJ, Girgis SI, MacIntyre I, Emson PC, Tohyama M (1987a) Calcitonin gene-related peptide-like immunoreactive sensory fibers form synaptic contact with sympathetic neurons in the rat celiac ganglion. Brain Res 407:149–151

Lee Y, Shiotani Y, Hayashi N, Kamada T, Hillyard CJ, Girgis SI, MacIntyre I, Tohyama M (1987b) Distribution and origin of calcitonin gene-related peptide in the rat stomach and duodenum: an immunohistochemical study. J Neural Transm 68:1–14

Le Grevès P, Nyberg F, Terenius L, Hökfelt T (1985) Calcitonin gene-related peptide is a potent inhibitor of substance P degradation. Eur J Pharmacol 115:309–311

Le Grevès P, Nyberg F, Hökfelt T, Terenius L (1989) Calcitonin gene-related peptide is metabolized by an endopeptidase hydrolyzing substance P. Regul Pept 25:277–286

Lembeck F (1983) Sir Thomas Lewis's nocifensor system, histamine and substance P-containing primary afferent nerves. Trends Neurosci 6:106–108

Lembeck F, Bucsics A (1990) Classification of peripheral neurones. Behav Brain Sci 13: 310–311

Lembeck F, Donnerer J (1981a) Postocclusive cutaneous vasodilatation mediated by substance P. Naunyn Schmiedebergs Arch Pharmacol 316:165–171

Lembeck F, Donnerer J (1981b) Time course of capsaicin-induced functional impairments in comparison with changes in neuronal substance P content. Naunyn Schmiedebergs Arch Pharmacol 316:240–243

Lembeck F, Donnerer J (1985) Opioid control of the function of primary afferent substance P fibres. Eur J Pharmacol 114:241–246

Lembeck F, Gamse R (1982) Substance P in peripheral sensory processes. In: Porter R, O'Connor M (eds) Substance P in the nervous system. Pitman, London, pp 35–49

Lembeck F, Holzbauer M (1988) Neuronal mechanisms of cutaneous blood flow. In: Burnstock G, Griffith SG (eds) Nonadrenergic innervation of blood vessels: II. Regional innervation. CRC Press, Boca Raton, pp 119–132

Lembeck F, Holzer P (1979) Substance P as neurogenic mediator of antidromic vasodilation and neurogenic plasma extravasation. Naunyn Schmiedebergs Arch Pharmacol 310: 175–183

Lembeck F, Skofitsch G (1982) Visceral pain reflex after pretreatment with capsaicin and morphine. Naunyn Schmiedebergs Arch Pharmacol 321:116–122

Lembeck F, Gamse R, Juan H (1977) Substance P and sensory nerve endings. In: von Euler US, Pernow B (eds) Substance P. Raven, New York, pp 169–181

Lembeck F, Folkers K, Donnerer J (1981) Analgesic effect of antagonists of substance P. Biochem Biophys Res Commun 103:1318–1321

Lembeck F, Donnerer J, Barthó L (1982) Inhibition of neurogenic vasodilatation and plasma extravasation by substance P antagonists, somatostatin and [D-Met^2,Pro^5]enkephalinamide. Eur J Pharmacol 85:171–176

Lembeck F, Donnerer J, Tsuchiya M, Nagahisa A (1992) The non-peptide tachykinin antagonist, CP-96,345, is a potent inhibitor of neurogenic inflammation. Br J Pharmacol 105:527–530

Leung FW (1992a) Modulation of autoregulatory escape by capsaicin-sensitive afferent nerves in the rat stomach. Am J Physiol 262:H562–H567

Leung FW (1992b) Role of capsaicin-sensitive afferent nerves in mucosal injury and injury-induced hyperemia in rat colon. Am J Physiol 262:G332–G337

Leung FW, Guth PH (1985) Dissociated effects of somatostatin on gastric acid secretion and mucosal blood flow. Am J Physiol 248:G337–G341

LeVasseur SA, Gibson SJ, Helme RD (1990) The measurement of capsaicin-sensitive sensory nerve fiber function in elderly patients with pain. Pain 41:19–25

Levine DJ, Clark R, Devor M, Helms C, Moskowitz MA, Basbaum AI (1984) Intraneuronal substance P contributes to the severity of experimental arthritis. Science 226:547–549

Levine DJ, Dardick SJ, Roizen MF, Helms C, Basbaum AI (1986) Contribution of sensory afferents and sympathetic efferents to joint injury in experimental arthritis. J Neurosci 6: 3423–3429

Lewis T (1927) The blood vessels of the human skin and their responses. Shaw and Sons, London

Lewis T (1937a) The nocifensor system of nerves and its reactions. Br Med J I:431–435

Lewis T (1937b) The nocifensor system of nerves and its reactions. Br Med J I:491–497

Lewis T, Marvin HM (1927) Observations relating to vasodilatation arising from antidromic impulses, to herpes zoster and trophic effects. Heart 14:27–46

Li D-S, Raybould HE, Quintero E, Guth PH (1991) Role of calcitonin gene-related peptide in gastric hyperemic response to intragastric capsaicin. Am J Physiol 261:G657–G661

Li D-S, Raybould HE, Quintero E, Guth PH (1992) Calcitonin gene-related peptide mediates the gastric hyperemic response to acid back-diffusion. Gastroenterology 102: 1124–1128

Li YJ, Duckles SP (1991a) Differential effects of neuropeptide Y and opioids on neurogenic responses of the perfused rat mesentery. Eur J Pharmacol 195:365–372

Li YJ, Duckles SP (1991b) Effect of opioid receptor antagonists on vasodilator nerve actions in the perfused rat mesentery. Eur J Pharmacol 204:323–328

Li YJ, Duckles SP (1992) Effect of endothelium on the actions of sympathetic and sensory nerves in the perfused rat mesentery. Eur J Pharmacol 210:23–30

Limlomwongse L, Chaitauchawong C, Tongyai S (1979) Effect of capsaicin on gastric acid secretion and mucosal blood flow. J Nutr 109:773–777

Lindgren BR, Anderson CD, Frödin T, Andersson RGG (1987) Inhibitory effects of clonidine on allergic reactions in guinea-pig skin. Eur J Pharmacol 134:339–343

Lindh B, Hökfelt T, Elfvin L-G, Terenius L, Fahrenkrug J, Elde R, Goldstein M (1986) Topography of NPY-, somatostatin- and VIP-immunoreactive, neuronal subpopulations in the guinea pig celiac-superior mesenteric ganglion and their projection to the pylorus. J Neurosci 6:2371–2383

Lindh B, Hökfelt T, Elfvin L-G (1988) Distribution and origin of peptide-containing nerve fibers in the celiac superior mesenteric ganglion of the guinea-pig. Neuroscience 26: 1037–1071

Linnik MD, Moskowitz MA (1989) Identification of immunoreactive substance P in human and other mammalian endothelial cells. Peptides 10:957–962

Lippe IT, Holzer P (1991) CGRP-induced mucosal hyperemia in the rat stomach is blocked by N^G-nitro-L-arginine methyl ester (L-NAME) and phenobarbital anesthesia. Regul Pept 34:122

Lippe IT, Holzer P (1992) Participation of endothelium-derived nitric oxide but not prostacyclin in the gastric mucosal hyperaemia due to acid back-diffusion. Br J Pharmacol 105:708–714

Lippe IT, Lorbach M, Holzer P (1989a) Close arterial infusion of calcitonin gene-related peptide into the rat stomach inhibits aspirin- and ethanol-induced hemorrhagic damage. Regul Pept 26:35–46

Lippe IT, Pabst MA, Holzer P (1989b) Intragastric capsaicin enhances rat gastric acid elimination and mucosal blood flow by afferent nerve stimulation. Br J Pharmacol 96: 91–100

Lippe IT, Stabentheiner A, Holzer P (1992) Nitric oxide participates in the vasodilator but not exudative component of mustard oil-induced inflammation in rat skin. Neuropeptides 22:41

Lisney SJW, Bharali LAM (1989) The axon reflex: an outdated idea or a valid hypothesis? News Physiol Sci 4:45–48

Loesch A, Burnstock G (1988) Ultrastructural localisation of serotonin and substance P in vascular endothelial cells of rat femoral and mesenteric arteries. Anat Embryol (Berl) 178:137–142

Longhurst JC, Ashton JH, Iwamoto GA (1980) Cardiovascular reflexes resulting from capsaicin-stimulated gastric receptors in anesthetized dogs. Circ Res 46:780–788

Lotz M, Vaughan JH, Carson DA (1988) Effect of neuropeptides on production of inflammatory cytokines by human monocytes. Science 241:1218–1221

Louis RE, Radermecker MF (1990) Substance P-induced histamine release from human basophils, skin and lung fragments: effect of nedrocomil sodium and theophylline. Int Arch Allergy Appl Immunol 92:329–333

Louis SM, Jamieson A, Russell NJW, Dockray GJ (1989a) The role of substance P and calcitonin gene-related peptide in neurogenic plasma extravasation and vasodilatation in the rat. Neuroscience 32:581–586

Louis SM, Johnstone D, Russell NJW, Jamieson A, Dockray GJ (1989b) Antibodies to calcitonin gene-related peptide reduce inflammation by topical mustard oil but not that due to carrageenin in the rat. Neurosci Lett 102:257–260

Louis SM, Johnstone D, Millest AJ, Russell NJW, Dockray GJ (1990) Immunization with calcitonin generelated peptide reduces the inflammatory response to adjuvant arthritis in the rat. Neuroscience 39:727–731

Low A, Westerman RA (1989) Neurogenic vasodilation in the rat hairy skin measured using a laser Doppler flowmeter. Life Sci 45:49–57

Lowman MA, Benyon RC, Church MK (1988) Characterization of neuropeptide-induced histamine release from human dispersed skin mast cells. Br J Pharmacol 95:121–130

Lundberg JM (1981) Evidence for coexistence ov vasoactive intestinal polypeptide (VIP) and acetylcholine in neurons of cat exocrine glands. Acta Physiol Scand 112 [Suppl 496]: 1–57

Lundberg JM, Saria A (1987) Polypeptide-containing neurons in airway smooth muscle. Annu Rev Physiol 49:557–572

Lundberg JM, Hökfelt T, Änggård A, Terenius L, Elde R, Markey K, Goldstein M, Kimmel J (1982) Organizational principles in the peripheral sympathetic nervous system: subdivision by coexisting peptides (somatostatin-, avian pancreatic polypeptide-, and vasoactive intestinal polypeptide-like immunoreactive materials). Proc Natl Acad Sci USA 79: 1303–1307

Lundberg JM, Brodin E, Hua X-Y, Saria A (1984a) Vascular permeability changes and smooth muscle contraction in relation to capsaicin-sensitive substance P afferents in the guinea-pig. Acta Physiol Scand 120:217–227

Lundberg JM, Saria A, Rosell S, Folkers K (1984b) A substance P antagonist inhibits heat-induced oedema in the rat skin. Acta Physiol Scand 120:145–146

Lundberg JM, Franco-Cereceda A, Hua X-Y, Hökfelt T, Fischer J (1985) Co-existence of substance P and calcitonin gene-related peptide immunoreactivities in sensory nerves in relation to cardiovascular and bronchoconstrictor effects of capsaicin. Eur J Pharmacol 108:315–319

Lundblad L, Lundberg JM, Änggård A, Zetterström D (1987) Capsaicin-sensitive nerves and the cutaneous allergy reaction in man. Possible involvement of sensory neuropeptides in the flare reaction. Allergy 42:20–25

Luthman J, Stromberg I, Brodin E, Jonsson G (1989) Capsaicin treatment to developing rats induces increase of noradrenaline levels in the iris without affecting the adrenergic terminal density. Int J Dev Neurosci 7:613–620

Lynn B (1988) Neurogenic inflammation. Skin Pharmacol 1:217–224

Lynn B, Cotsell B (1991) The delay in onset of vasodilator flare in human skin at increasing distances from a localized noxious stimulus. Microvasc Res 41:197–202

Lynn B, Shakhanbeh J (1988a) Neurogenic inflammation in the skin of the rabbit. Agents Actions 25:228–230

Lynn B, Shakhanbeh J (1988b) Substance P content of the skin, neurogenic inflammation and numbers of C-fibres following capsaicin application to a cutaneous nerve in the rabbit. Neuroscience 24:769–776

MacNaughton WK, Cirino G, Wallace JL (1989) Endothelium-derived relaxing factor (nitric oxide) has protective actions in the stomach. Life Sci 45:1869–1876

Magerl W, Szolcsányi J, Westerman RA, Handwerker HO (1987) Laser Doppler measurements of skin vasodilation elicited by percutaneous electrical stimulation of nociceptors in humans. Neurosci Lett 82:349–354

Maggi CA (1991) The pharmacology of the efferent function of sensory nerves. J Auton Pharmacol 11:173–208

Maggi CA, Meli A (1988) The sensory-efferent function of capsaicin-sensitive sensory neurons. Gen Pharmacol 19:1–43

Maggi CA, Giuliani S, Santicioli P, Regoli D, Meli A (1985) Comparison of the effects of substance P and substance K on blood pressure, salivation and urinary bladder motility in urethane-anaesthetized rats. Eur J Pharmacol 113:291–294

Maggi CA, Borsini F, Santicioli P, Geppetti P, Abelli L, Evangelista S, Manzini S, Theodorsson-Norheim E, Somma V, Amenta F, Bacciarelli C, Meli A (1987a) Cutaneous lesions in capsaicin-pretreated rats. A trophic role of capsaicin-sensitive afferents? Naunyn Schmiedebergs Arch Pharmacol 336:538–545

Maggi CA, Evangelista S, Abelli L, Somma V, Meli A (1987b) Capsaicin-sensitive mechanisms and experimentally induced duodenal ulcers in rats. J Pharm Pharmacol 39: 559–561

Maggi CA, Evangelista S, Giuliani S, Meli A (1987c) Anti-ulcer activity of calcitonin gene-related peptide in rats. Gen Pharmacol 18:33–34

Maggi CA, Santicioli P, Del Bianco E, Geppetti P, Barbanti G, Turini D, Meli A (1989a) Release of VIP- but not CGRP-like immunoreactivity by capsaicin from the human isolated small intestine. Neurosci Lett 98:317–320

Maggi CA, Santicioli P, Renzi D, Patacchini R, Surrenti C, Meli A (1989b) Release of substance P- and calcitonin gene-related peptide-like immunoreactivity and motor response of the isolated guinea pig gallbladder to capsaicin. Gastroenterology 96:1093–1101

Maggi CA, Patacchini R, Giuliani S, Rovero P, Dion S, Regoli D, Giachetti A, Meli A (1990) Competitive antagonists discriminate between NK_2 tachykinin receptor subtypes. Br J Pharmacol 100:588–592

Maggio JE, Hunter JC (1984) Regional distribution of kassinin-like immunoreactivity in rat central and peripheral tissues and the effect of capsaicin. Brain Res 307:370–373

Majno G, Shea SM, Leventhal M (1969) Endothelial contraction induced by histamine-type mediators. An electron microscopic study. J Cell Biol 42:647–670

Manela FD, Young RL, Ren R, Harty RF (1991) Stimulation of rat antral CGRP release by intraluminal peptone. Regul Pept 34:119

Mantione CR, Rodriguez R (1990) A bradykinin $(BK)_1$ receptor antagonist blocks capsaicin-induced ear inflammation in mice. Br J Pharmacol 99:516–518

Mantyh CR, Gates TS, Zimmerman RP, Welton ML, Passaro EP, Vigna SR, Maggio JE, Kruger L, Mantyh PW (1988a) Receptor binding sites for substance P, but not substance K or neuromedin K, are expressed in high concentrations by arterioles, venules, and lymph nodules in surgical specimens obtained from patients with ulcerative colitis and Crohn disease. Proc Natl Acad Sci USA 85:3235–3239

Mantyh PW, Mantyh CR, Gates T, Vigna SR, Maggio JE (1988b) Receptor binding sites for substance P and substance K in the canine gastrointestinal tract and their possible role in inflammatory bowel disease. Neuroscience 25:817–837

Mantyh PW, Catton MD, Boehmer CG, Welton ML, Passaro EP, Maggio JE, Vigna SR (1989) Receptors for sensory neuropeptides in human inflammantory diseases: implications for the effector role of sensory neurons. Peptides 10:627–645

Manzini S, Perretti F (1988) Vascular effects of capsaicin in isolated perfused rat mesenteric bed. Eur J Pharmacol 148:153–159

Manzini S, Perretti F, Tramontana M, Del Bianco E, Santicioli P, Maggi CA, Geppetti P (1991) Neurochemical evidence of calcitonin gene-related peptide-like immunoreactivity (CGRP-LI) release from capsaicin-sensitive nerves in rat mesenteric arteries and veins. Gen Pharmacol 22:275–278

Marabini S, Ciabatti G, Polli G, Fusco BM, Geppetti P, Maggi CA, Fanciullacci M, Sicuteri F (1988) Effect of topical nasal treatment with capsaicin in vasomotor rhinitis. Regul Pept 22:121

Marshall I, Al-Kazwini SJ, Holman JJ, Craig RK (1986) Human and rat alpha-CGRP but not calcitonin cause mesenteric vasodilatation in rats. Eur J Pharmacol 123:217–222

Martin HA, Basbaum AI, Kwiat GC, Goetzl EJ, Levine JD (1987) Leukotriene and prostaglandin sensitization of cutaneous high threshold C- and A-delta mechanociceptors in the hairy skin of rat hindlimbs. Neuroscience 22:651–659

Martins MA, Shore SA, Drazen JM (1991) Capsaicin-induced release of tachykinins: effects of enzyme inhibitors. J Appl Physiol 70:1950–1956

Martinson J (1965) Studies on the vagal efferent control of the stomach. Acta Physiol Scand 65 [Suppl 255]:1–24

Martling C-R, Lundberg JM (1988) Capsaicin-sensitive afferents contribute to the acute airway edema following tracheal instillation of hydrochloric acid or gastric juice in the rat. Anesthesiology 68:350–356

Maton PN, Sutliff VE, Zhou Z-C, Collins SM, Gardner JD, Jensen RT (1988) Characterization of receptors for calcitonin gene-related peptide on gastric smooth muscle cells. Am J Physiol 254:G789–G794

Maton PN, Pradhan T, Zhou Z-C, Gardner JD, Jensen RT (1990) Activities of calcitonin gene-related peptide (CGRP) and related peptides at the CGRP receptor. Peptides 11: 485–489

Matsas R, Fulcher IS, Kenny AJ, Turner AJ (1983) Substance P and [Leu]-enkephalin are hydrolysed by an enzyme in pig caudate synaptic membranes that is identical with the endopeptidase of kidney microvilli. Proc Natl Acad Sci USA 80:3111–3115

Matsumoto J, Takeuchi K, Okabe S (1991a) Characterization of gastric mucosal blood flow response induced by intragastric capsaicin in rats. Jpn J Pharmacol 57:205–213

Matsumoto J, Ueshima K, Takeuchi K, Okabe S (1991b) Capsaicin-sensitive afferent neurons in adaptive responses of the rat stomach induced by a mild irritant. Jpn J Pharmacol 55:181–185

Matsuyama T, Wanaka A, Yoneda S, Kimura K, Kamada T, Girgis S, MacIntyre I, Emson PC, Tohyama M (1986) Two distinct calcitonin gene-related peptide-containing peripheral nervous systems: distribution and quantitative differences between the iris and cerebral artery with special reference to substance P. Brain Res 373:205–212

Matthews B (1976) Coupling between cutaneous nerves. J Physiol (Lond) 254:37P–38P

Matthews MR, Cuello AC (1984) The origin and possible significance of substance P immunoreactive networks in the prevertebral ganglia and related structures in the guinea-pig. Philos Trans R Soc Lond [Biol] 306:247–276

Mayer EA, Baldi JP (1991) Can regulatory peptides be regarded as words of a biological language? Am J Physiol 261:G171–G184

Mayer EA, Koelbel CBM, Snape WJ, Eysselein V, Ennes H, Kodner A (1990) Substance P and CGRP mediate motor response of rabbit colon to capsaicin. Am J Physiol 259: G889–G897

Maynard KI, Saville VL, Burnstock G (1990) Sensory-motor neuromodulation of sympathetic vasoconstriction in the rabbit central ear artery. Eur J Pharmacol 187:171–182

Mazurek N, Pecht I, Teichberg VI, Blumberg S (1981) The role of the N-terminal tetrapeptide in the histamine releasing action of substance P. Neuropharmacology 20:1025–1027

McCarthy PW, Lawson SN (1989) Cell type and conduction velocity of rat primary sensory neurons with substance P-like immunoreactivity. Neuroscience 28:745–753

McCarthy PW, Lawson SN (1990) Cell type and conduction velocity of rat primary sensory neurons with calcitonin gene-related peptide-like immunoreactivity. Neuroscience 34: 623–632

McCusker MT, Chung KF, Roberts NM, Barnes PJ (1989) Effect of topical capsaicin on the cutaneous responses to inflammatory mediators and to antigen in man. J Allergy Clin Immunol 83:1118–1123

McGillis JP, Mitsuhashi M, Payan DG (1990) Immunomodulation by tachykinin neuropeptides. Ann NY Acad Sci 594:85–94

McGregor GP, Conlon JM (1991) Regulatory peptide and serotonin content and brush-border enzyme activity in the rat gastrointestinal tract following neonatal treatment with capsaicin; lack of effect on epithelial markers. Regul Pept 32:109–119

McKnight AT, Maguire JJ, Elliot NJ, Fletcher AE, Foster AC, Tridgett R, Williams BJ, Longmore J, Iversen LL (1991) Pharmacological specificity of novel, synthetic, cyclic peptides as antagonists at tachykinin receptors. Br J Pharmacol 104:355–360

McMahon S, Koltzenburg M (1990) The changing role of primary afferent neurones in pain. Pain 43:269–272

McMahon SB, Lewin GR, Anand P, Ghatei MA, Bloom SR (1989) Quantitative analysis of peptide levels and neurogenic extravasation following regeneration of afferents to appropriate and inappropriate targets. Neuroscience 33:67–73

Meehan AG, Hottenstein OD, Kreulen DL (1991) Capsaicin-sensitive nerves mediate inhibitory junction potentials and dilatation in guinea-pig mesenteric artery. J Physiol (Lond) 443:161–174

Melchiorri P, Tonelli F, Negri L (1977) Comparative circulatory effects of substance P, eledoisin and physalaemin in the dog. In: von Euler US, Pernow B (eds) Substance P. Raven, New York, pp 311–319

Meyer RA, Raja SN, Campbell JN (1985) Coupling of action potential activity between unmyelinated fibers in the peripheral nerve of monkey. Science 227:184–187

Mimeault M, Fournier A, Dumont Y, St Pierre S, Quirion R (1991) Comparative affinities and antagonistic potencies of various human calcitonin gene-related peptide fragments on calcitonin gene-related peptide receptors in brain and periphery. J Pharmacol Exp Ther 258:1084–1090

Minagawa H, Shiosaka S, Inoue H, Hayashi N, Kasahara A, Kamada T, Tohyama M, Shitani Y (1984) Origins and three-dimensional distribution of substance P-containing structures in the rat stomach using whole-mount tissue. Gastroenterology 86:51–59

Molander C, Ygge J, Dalsgaard J-C (1987) Substance P-, somatostatin- and calcitonin gene-related peptide-like immunoreactivity and fluoride resistant acid phosphatase-activity in relation to retrogradely labeled cutaneous, muscular and visceral primary sensory neurons in the rat. Neurosci Lett 74:37–42

Moncada S, Palmer RMJ, Higgs EA (1991) Nitric oxide: physiology, pathophysiology, and pharmacology. Pharmacol Rev 43:109–142

Morishita T, Guth PH (1986) Vagal nerve stimulation causes noncholinergic dilatation of gastric arterioles. Am J Physiol 250:G660–G664

Moritoki H, Takase H, Tanioka A (1990) Dual effects of capsaicin on responses of the rabbit ear artery to field stimulation. Br J Pharmacol 99:152–156

Morton CR, Chahl LA (1980) Pharmacology of the neurogenic oedema response to electrical stimulation of the saphenous nerve in the rat. Naunyn Schmiedebergs Arch Pharmacol 314:271–276

Moskowitz MA, Buzzi MG, Sakas DE, Linnik MD (1989) Pain mechanisms underlying vascular headaches. Rev Neurol (Paris) 145:181–193

Mulderry PK, Ghatei MA, Rodrigo J, Allen JM, Rosenfeld MG, Polak JM, Bloom SR (1985) Calcitonin gene-related peptide in cardiovascular tissues of the rat. Neuroscience 14:947–954

Mulderry PK, Ghatei MA, Spokes RA, Jones PM, Pierson AM, Hamid QA, Kanse S, Amara SG, Burrin JM, Legon S, Polak JM, Bloom SR (1988) Differential expression of α-CGRP and β-CGRP by primary sensory neurons and enteric autonomic neurons of the rat. Neuroscience 25:195–205

Nadel JA (1992) Regulation of neurogenic inflammation by neutral endopeptidase. Am Rev Respir Dis 145:S48–S52

Nagy JI, Vincent SR, Staines WMA, Fibiger HC, Reisine TD, Yamamura HI (1980) Neurotoxic action of capsaicin on spinal substance P neurons. Brain Res 186:435–444

Nagy JI, Hunt SP, Iversen LL, Emson PC (1981) Biochemical and anatomical observations on the degeneration of peptide-containing afferent neurons after neonatal capsaicin. Neuroscience 6:1923–1934

Nakanishi S (1987) Substance P precursor and kininogen: their structures, gene organizations, and regulation. Physiol Rev 67:1117–1142

Naukkarinen A, Harvima IT, Aalto ML, Harvima RJ, Horsmanheimo M (1991) Quantitative analysis of contact sites between mast cells and sensory nerves in cutaneous psoriasis and lichen planus based on a histochemical double staining technique. Arch Dermatol Res 283:433–437

Nelson MT, Huang Y, Brayden JE, Hescheler J, Standen NB (1990) Arterial dilations in response to calcitonin gene-related peptide involve activation of K^+ channels. Nature 344:770–773

Neuhuber WL (1987) Sensory vagal innervation of the rat esophagus and cardia: a light and electron microscopic anterograde tracing study. J Auton Nerv Syst 20:243–255

Nielsch U, Keen P (1987) Effects of neonatal 6-hydroxydopamine administration on different substance P-containing sensory neurones. Eur J Pharmacol 138:193–197

Nilsson G (1989) Modulation of the immune response in rat by utilizing the neurotoxin capsaicin. Thesis, University of Uppsala

Nilsson J, von Euler AM, Dalsgaard C-J (1985) Stimulation of connective tissue cell growth by substance P and substance K. Nature 315:61–63

Nilsson J, Sejersen T, Nilsson AH, Dalsgaard C-J (1986) DNA synthesis induced by the neuropeptide substance K correlates to the level of myc-gene transcripts. Biochem Biophys Res Commun 137:167–174

Noguchi K, Senba E, Morita Y, Sato M, Tohyama M (1990) Co-expression of α-CGRP and β-CGRP mRNAs in the rat dorsal root ganglion cells. Neurosci Lett 108:1–5

Nordin M (1990) Low-threshold mechanoreceptive and nociceptive units with unmyelinated (C) fibres in the human supraorbital nerve. J Physiol (Lond) 426:229–240

Oates PJ (1990) Gastric blood flow and mucosal defense. In: Hollander D, Tarnawski AS (eds) Gastric cytoprotection. Plenum, New York, pp 125–165

O'Brien C, Woolf CJ, Fitzgerald M, Lindsay RM, Molander C (1989) Differences in the chemical expression of rat primary afferent neurons which innervate skin, muscle or joint. Neuroscience 32:493–502

O'Flynn NM, Helme RD, Watkins DJ, Burcher E (1989) Autoradiographic localization of substance P binding sites in rat footpad skin. Neurosci Lett 106:43–48

Ohkubo N, Nakanishi S (1991) Molecular characterization of the three tachykinin receptors. Ann NY Acad Sci 632:53–62

Öhlén A, Wiklund NP, Persson MG, Hedqvist P (1988) Calcitonin gene-related peptide desensitizes skeletal muscle arterioles to substance P in vivo. Br J Pharmacol 95:673–674

Öhlén A, Thureson-Klein A, Lindbom L, Persson MG, Hedqvist P (1989) Substance P activates leukocytes and platelets in rabbit microvessels. Blood Vessels 26:84–94

Olesen J, Edvinsson L (1991) Migraine: a research field matured for the basic neurosciences. Trends Neurosci 14:3–5

Ordway GA, Longhurst JC (1983) Cardiovascular reflexes arising from the gallbladder of the cat. Effects of capsaicin, bradykinin, and distension. Circ Res 52:26–35

Ordway GA, Boheler KR, Longhurst JC (1988) Stimulating intestinal afferents reflexly activates cardiovasular system in cats. Am J Physiol 254:H354–H360

Osswald W (1990) Vascular hypertrophy caused by sympathetic denervation. Eur J Pharmacol 183:89

Papka RE, Furness JB, Della NG, Murphy R, Costa M (1984) Time course of effect of capsaicin on ultrastructure and histochemistry of substance P-immunoreactive nerves associated with the cardiovascular system of the guinea-pig. Neuroscience 12:1277–1292

Parkman HP, Reynolds JC, Elfman KS, Ogorek CP (1989) Calcitonin gene-related peptide: a sensory and motor neurotransmitter in the feline lower esophageal sphincter. Regul Pept 25:131–146

Parrot JL (1942) Les modifications apportées par un antagoniste de l'histamine (2339 RP) à la réaction vasculaire locale de la peau. C R Soc Biol (Paris) 136:715–716

Patacchini R, Astolfi M, Quartara L, Rovero P, Giachetti A, Maggi CA (1991) Further evidence for the existence of NK_2 tachykinin receptor subtypes. Br J Pharmacol 104:91–96

Payan DG (1985) Receptor-mediated mitogenic effects of substance P on cultured smooth muscle cells. Biochem Biophys Res Commun 130:104–109

Payan DG, Levine JD, Goetzl EJ (1984) Modulation of immunity and hypersensitivity by sensory neuropeptides. J Immunol 132:1601–1604

Pearce FL, Kassessinoff TA, Liu WL (1989) Characteristics of histamine secretion induced by neuropeptides: implications for the relevance of peptide-mast cell interactions in allergy and inflammation. Int Arch Allergy Appl Immunol 88:129–131

Pedersen-Bjergaard U, Nielsen LB, Jensen K, Edvinsson L, Jansen I, Olesen J (1989) Algesia and local responses induced by neurokinin A and substance P in human skin and temporal muscle. Peptides 10:1147–1152

Pedersen-Bjergaard U, Nielsen LB, Jensen K, Edvinsson L, Jansen I, Olesen J (1991) Calcitonin gene-related peptide, neurokinin A and substance P: effects on nociception and neurogenic inflammation in human skin and temporal muscle. Peptides 12:333–337

Pernow B (1983) Substance P. Pharmacol Rev 35:85–141

Pernow J (1989) Actions of constrictor (NPY and endothelin) and dilator (substance P, CGRP and VIP) peptides on pig splenic and human skeletal muscle arteries: involvement of the endothelium. Br J Pharmacol 97:983–989

Peskar BM, Respondek M, Müller KM, Peskar BA (1991) A role for nitric oxide in capsaicin-induced gastroprotection. Eur J Pharmacol 198:113–114

Petersen BM, Trad KS, Hakki FA, Harmon JW, Bass BL (1991a) Capsaicin-induced augmentation of esophageal mucosal blood flow is independent of cyclooxygenase blockade. Gastroenterology 100:A139

Petersen BM, Trad KS, Hakki FA, Harmon JW, Bass BL (1991b) Esophageal mucosal blood flow augmentation induced by luminal capsaicin is ablated by capsaicin treatment of vagal and spinal afferents. Gastroenterology 100:A139

Pierau F-K, Szolcsányi J (1989) Neurogenic inflammation: axon reflex in pigs. Agents Actions 26:231–232

Pierau F-K, Ewert C, Ernst R, Barthó L, Sann H (1991) Flare reaction and plasma extravasation in the skin of pigs. Abstract of the 1st international meeting of the European Neuropeptide Club: functional and pharmacological aspects of neuropeptides, Igls/Innsbruck, p 79

Piotrowski W, Foreman JC (1985) On the actions of substance P, somatostatin, and vasoactive intestinal polypeptide on rat peritoneal mast cells and in human skin. Naunyn Schmiedebergs Arch Pharmacol 331:364–368

Piotrowski W, Foreman JC (1986) Some effects of calcitonin gene-related peptide in human skin and on histamine release. Br J Dermatol 114:37–46

Piotrowski W, Devoy MAB, Jordan CC, Foreman JC (1984) The substance P receptor on rat mast cells and in human skin. Agents Actions 14:420–424

Pique JM, Esplugues JV, Whittle BJR (1989) The vasodilator role of endogenous nitric oxide in the rat gastric microcirculation. Eur J Pharmacol 174:293–296

Pique JM, Esplugues JV, Whittle BJR (1990) Influence of morphine or capsaicin pretreatment on rat gastric microcirculatory response to PAF. Am J Physiol 258:G352–G357

Pitetti KH, Cole DJ, Ordway GA (1988) Activation of gastric afferent fibers increases coronary arterial resistance in anesthetized dogs. J Auton Nerv Syst 23:25–34

Pohl U (1990) Endothelial cells as part of a vascular oxygen-sensing system: hypoxia-induced release of autacoids. Experientia 46:1175–1179

Prechtl JC, Powley TL (1990) B-afferents: a fundamental division of the nervous system mediating homeostasis? Behav Brain Sci 13:289–331

Ralevic V, Milner P, Hudlická O, Kristek F, Burnstock G (1990) Substance P is released from the endothelium of normal and capsaicin-treated rat hind-limb vasculature, in vivo, by increased flow. Circ Res 66:1178–1183

Ralevic V, Khalil Z, Dusting GJ, Helme RD (1992) Nitric oxide and sensory nerves are involved in the vasodilator response to acetylcholine but not calcitonin gene-related peptide in rat skin microvasculature. Br J Pharmacol 106:650–655

Rang HP, Bevan S, Dray A (1991) Chemical activation of nociceptive peripheral neurones. Br Med Bull 47:534–548

Raud J, Lundeberg T, Brodda-Jansen G, Theodorsson E, Hedqvist P (1991) Potent anti-inflammatory action of calcitonin gene-related peptide. Biochem Biophys Res Commun 180:1429–1435

Raybould HE, Taché Y (1988) Cholecystokinin inhibits gastric motility and emptying via a capsaicin-sensitive vagal pathway in rats. Am J Physiol 255:G242–G246

Raybould HE, Taché Y (1989) Capsaicin-sensitive vagal afferent fibers and stimulation of gastric acid secretion in anesthetized rats. Eur J Pharmacol 167:237–243

Raybould HE, Holzer P, Reddy SN, Yang H, Taché Y (1990) Capsaicin-sensitive vagal afferents contribute to gastric acid and vascular responses to intracisternal TRH analog. Peptides 11:789–795

Raybould HE, Sternini C, Eysselein VE, Yoneda M, Holzer P (1992) Selective ablation of spinal afferent neurons containing CGRP attenuates gastric hyperemic response to acid. Peptides 113:249–254

Regoli D, D'Orléans-Juste P, Escher E, Mizrahi J (1984) Receptors for substance P: I. The pharmacological preparations. Eur J Pharmacol 97:161–170

Regoli D, Drapeau G, Dion S, D'Orléans-Juste PD (1989) Receptors for substance P and related neurokinins. Pharmacology 38:1–15

Remak G, Hottenstein OD, Jacobson ED (1990) Sensory nerves mediate neurogenic escape in rat gut. Am J Physiol 258:H778–H786

Ren J, Young RL, Lassiter DC, Harty RF (1991) Effects of hCGRP receptor antagonist on capsaicin-mediated release of regulatory peptides and acetylcholine from rat antrum. Gastroenterology 100:A661

Renzi D, Santicioli P, Maggi CA, Surrenti C, Pradelles P, Meli A (1988) Capsaicin-induced release of substance P-like immunoreactivity from the guinea pig stomach in vitro and in vivo. Neurosci Lett 92:254–258

Renzi D, Evangelista S, Mantellini P, Santicioli P, Maggi CA, Geppetti P, Surrenti C (1991) Capsaisin-induced release of neurokinin A from muscle and mucosa of gastric corpus: correlation with capsaicin-evoked release of calcitonin gene-related peptide. Neuropeptides 19:137–145

Ritchie WP (1975) Acute gastric mucosal damage induced by bile salts, acid and ischaemia. Gastroenterology 68:699–707

Robert A, Olafsson AS, Lancaster C, Gilbertson-Beadling S (1990) Acute capsaicin inhibits aspirin ulcers and acid secretion in rats, but capsaicin denervation has no effect. Gastroenterology 98:A113

Rodrigo J, Polak JM, Fernandez L, Ghatei MA, Mulderry P, Bloom SR (1985) Calcitonin gene-related peptide immunoreactive sensory and motor nerves of the rat, cat, and monkey esophagus. Gastroenterology 88:444–451

Rodrigue F, Hoff P, Touvay C, Vilain B, Carré C, Mencia-Huerta J-M, Braquet P (1988) Platelet-activating factor induces the release of substance P and vasoactive intestinal peptide from guinea pig lung tissues. In: Braquet P (ed) The role of platelet-activating factor in immune disorders. Karger, Basel, pp 93–98

Rosell S, Olgart L, Gazelius B, Panopoulos P, Folkers K, Hörig J (1981) Inhibition of antidromic and substance P-induced vasodilatation by a substance P antagonist. Acta Physiol Scand 111:381–382

Rosenfeld MG, Mermod J-J, Amara SG, Swanson LW, Sawchenko PE, Rivier J, Vale WW, Evans RM (1983) Production of a novel neuropeptide encoded by the calcitonin gene via tissue-specific RNA processing. Nature 304:129–135

Rouissi N, Nantel F, Drapeau G, Rhaleb N-E, Dion S, Regoli D (1990) Inhibitors of peptidases: how they influence the biological activities of substance P, neurokinins, kinins and angiotensins in isolated vessels. Pharmacology 40:185–195

Rovero P, Pestellini V, Patacchini R, Giuliani S, Santicioli P, Maggi CA, Meli A, Giachetti A (1989) A potent and selective agonist for NK_2 tachykinin receptor. Peptides 10: 593–595

Rózsa Z, Jacobson ED (1989) Capsaicin-sensitive nerves are involved in bile-oleate-induced intestinal hyperemia. Am J Physiol 256:G476–G481

Rózsa Z, Jancsó G, Varró V (1984) Possible involvement of capsaicin-sensitive sensory nerves in the regulation of intestinal blood flow in the dog. Naunyn Schmiedebergs Arch Pharmacol 326:352–356

Rózsa Z, Varró V, Jancsó G (1985) Use of immunoblockade to study the involvement of peptidergic afferent nerves in the intestinal vasodilatory response to capsaicin in the dog. Eur J Pharmacol 115:59–64

Rózsa Z, Sharkey KA, Jancsó G, Varró V (1986) Evidence for a role of capsaicin-sensitive mucosal afferent nerves in the regulation of mesenteric blood flow in the dog. Gastroenterology 90:906–910

Rózsa Z, Mattila J, Jacobson ED (1988) Substance P mediates a gastrointestinal thermoreflex in rats. Gastroenterology 95:265–276

Rubinstein I, Iwamoto I, Ueki IF, Borson DB, Nadel JA (1990) Recombinant neutral endopeptidase attenuates substance P-induced plasma extravasation in guinea pig skin. Int Arch Allergy Appl Immunol 91:232–238

Ruff MR, Wahl SM, Pert CP (1985) Substance P receptor-mediated chemotaxis of human monocytes. Peptides 6 [Suppl 2]:107–111

Said SI, Mutt V (1970) Potent peripheral and splanchnic vasodilator peptide from normal gut. Nature 225:863–864

Sakas DE, Moskowitz MA, Wei EP, Kontos HA, Kano M, Ogilvy CS (1989) Trigeminovascular fibers increase blood flow in cortical gray matter by axon reflex-like mechanisms during acute severe hypertension or seizures. Proc Natl Acad Sci USA 86: 1401–1405

Samuelson UE, Jernbeck J (1991) Calcitonin gene-related peptide relaxes porcine arteries via one endothelium-dependent and one endothelium-independent mechanism. Acta Physiol Scand 141:281–282

Sann H, Pintér E, Szolcsányi J, Pierau F-K (1988) Peptidergic afferents might contribute to the regulation of the skin blood flow. Agents Actions 23:14–15

Saria A (1984) Substance P in sensory nerve fibres contributes to the development of oedema in the rat hind paw after thermal injury. Br J Pharmacol 82:217–222

Saria A, Wolf G (1988) Beneficial effect of topically applied capsaicin in the treatment of hyperreactive rhinopathy. Regul Pept 22:167

Saria A, Lundberg JM, Skofitsch G, Lembeck F (1983) Vasuclar protein lekeage in various tissues induced by substance P, capsaicin, bradykinin, serotonin, histamine and by antigen challenge. Naunyn Schmiedebergs Arch Pharmacol 324:212–218

Saria A, Hua X-Y, Skofitsch G, Lundberg JM (1984) Inhibition of compound 48/80-induced vascular protein leakage by pretreatment with capsaicin and a substance P antagonist. Naunyn Schmiedebergs Arch Pharmacol 328:9–15

Sayadi H, Harmon JW, Moody TW, Korman LY (1988) Autoradiographic distribution of vasoactive intestinal polypeptide receptors in rabbit and rat small intestine. Peptides 9: 23–30

Schaible H-G, Schmidt RF (1988) Time course of mechanosensitivity changes in articular afferents during a developing arthritis. J Neurophysiol 60:2180–2195

Schini L, Katusic ZS, Vanhoutte PM (1990) Neurohypophyseal peptides and tachykinins stimulate the production of cyclic GMP in cultured porcine aortic endothelial cells. J Pharmacol Exp Ther 255:994–1000

Schrauwen E, Houvenaghel A (1980) Substance P: a powerful intestinal vasodilator in the pig. Pflügers Arch 386:281–284

Sekizawa K, Tamaoki J, Graf PD, Basbaum CB, Borson DB, Nadel JA (1987) Enkephalinase inhibitor potentiates mammalian tachykinin-induced contraction in ferret trachea. J Pharmacol Ex Ther 243:1211–1217

Sestini P, Bienenstock J, Crowe SE, Marshall JS, Stead RH, Kakuta Y, Perdue MH (1990) Ion transport in rat tracheal epithelium in vitro. Role of capsaicin-sensitive nerves in allergic reactions. Am Rev Respir Dis 141:393–397

Shanahan F, Denburg JA, Fox J, Bienenstock J, Befus D (1985) Mast cell heterogeneity: effects of neuroenteric peptides on histamine release. J Immunol 135:1331–1337

Sharkey KA, Williams RG, Dockray GJ (1984) Sensory substance P innervation of the stomach and pancreas. Demonstration of capsaicin-sensitive sensory neurons in the rat by combined immunohistochemistry and retrograde tracing. Gastroenterology 87:914–921

Shibata M, Ohkubo T, Takahashi H, Inoki R (1986) Interaction of bradykinin with substance P on vascular permeability and pain response. Jpn J Pharmacol 41:427–429

Shimizu T, Fujita S, Izumi K, Koja T, Ohba N, Fukuda T (1984) Corneal lesions induced by the systemic administration of capsaicin in neonatal mice and rats. Naunyn Schmiedebergs Arch Pharmacol 326:347–351

Shimizu T, Izumi K, Fujita S, Koja T, Sorimachi M, Ohba N, Fukuda T (1987) Capsaicin-induced corneal lesions in mice and the effects of chemical sympathectomy. J Pharmacol Exp Ther 243:690–695

Shoji T, Ishira H, Ishikawa T, Saito A, Goto K (1987) Vasodilating effects of human and rat calcitonin gene-related peptides in isolated porcine coronary arteries. Naunyn Schmiedebergs Arch Pharmacol 336:438–444

Sidawy AN, Sayadi H, Harmon JW, Termanini B, Andrews B, DePalma RG, Korman LY (1989) Distribution of vasoactive intestinal peptide and its receptors in the arteries of the rabbit. J Surg Res 47:105–111

Simone DA, Ochoa J (1991) Early and late effects of prolonged topical capsaicin on cutaneous sensibility and neurogenic vasodilatation in humans. Pain 47:285–294

Skofitsch G, Jacobowitz DM (1985a) Galanin-like immunoreactivity in capsaicin-sensitive sensory neurons and ganglia. Brain Res Bull 15:191–195

Skofitsch G, Jacobowitz DM (1985b) Calcitonin gene-related peptide coexists with substance P in capsaicin sensitive neurons and sensory ganglia of the rat. Peptides 6: 747–754

Skofitsch G, Donnerer J, Petronijevic S, Saria A, Lembeck F (1983) Release of histamine by neuropeptides from the perfused rat hindquarter. Naunyn Schmiedebergs Arch Pharmacol 322:153–157

Skofitsch G, Savitt JM, Jacobowitz DM (1985a) Suggestive evidence for a functional unit between mast cells and substance P fibers in the rat diaphragm and mesentery. Histochemistry 82:5–8

Skofitsch G, Zamir N, Helke CJ, Savitt JM, Jacobowitz DM (1985b) Corticotropin releasing factor-like immunoreactivity in sensory ganglia and capsaicin sensitive neurons of the rat central nervous system: colocalization with other neuropeptides. Peptides 6: 307–318

Smith TW, Buchan P (1984) Peripheral opioid receptors located on the rat saphenous nerve. Neuropeptides 5:217–220

Snider RM, Constantine JW, Lowe JA, Longo KP, Lebel WS, Woody HA, Drozda SE, Desai MC, Vinick FJ, Spencer RW, Hess H-J (1991) A potent nonpeptide antagonist of substance P (NK_1) receptor. Science 251: 435–437

Spina D, McKenniff MG, Coyle AJ, Seeds EAM, Tramontana M, Peretti F, Manzini S, Page CP (1991) Effect of capsaicin on PAF-induced bronchial hyperresponsiveness and pulmonary cell accumulation in the rabbit. Br J Pharmacol 103:1268–1274

Starke K (1964) Substanz P-Gehalt und permeabilitätserhöhende Wirkung in verschiedenen Gehirngebieten. Naunyn Schmiedebergs Arch Exp Pathol Pharmakol 247:149–163

Starlinger M, Schiessel R, Hung CR, Silen W (1981) H^+ back diffusion stimulating gastric mucosal blood flow in the rabbit fundus. Surgery 89:232–236

Staszewska-Woolley J, Luk DE, Nolan PN (1986) Cardiovascular reflexes mediated by capsaicin sensitive cardiac afferent neurones in the dog. Cardiovasc Res 20:897–906

Stead RH, Bienenstock J, Stanisz AM (1987a) Neuropeptide regulation of mucosal immunity. Immunol Rev 100:333–359

Stead RH, Tomioka M, Quinonez G, Simon GT, Felten SY, Bienenstock J (1987b) Intestinal mucosal mast cells in normal and nematode-infected rat intestines are in intimate contact with peptidergic nerves. Proc Natl Acad Sci USA 84:2975–2979

Stead RH, Dixon MF, Bramwell NH, Riddel RH, Bienenstock J (1989) Mast cells are closely apposed to nerves in the human gastrointestinal mucosa. Gastroenterology 97: 575–585

Steen KH, Reeh PW, Anton F, Handwerker HO (1992) Protons selectively induce lasting excitation and sensitization to mechanical stimulation of nociceptors in rat skin in vitro. J Neurosci 12:86–95

Stein C, Hassan AHS, Przewlocki R, Gramsch C, Peter K, Herz A (1990) Opioids from immunocytes interact with receptors on sensory nerves to inhibit nociception in inflammation. Proc Natl Acad Sci USA 87:5935–5939

Sternini C (1991) Tachykinin and calcitonin gene-related peptide immunoreactivities and mRNAs in the mammalian enteric nervous system and sensory ganglia. In: Costa M, Surrenti C, Gorini S, Maggi CA, Meli A (eds) Sensory nerves and neuropeptides in gastroenterology. Plenum, New York, pp 39–51

Sternini C, Reeve JR, Brecha N (1987) Distribution and characterization of calcitonin gene-related peptide immunoreactivity in the digestive system of normal and capsaicin-treated rats. Gastroenterology 93:852–862

Sternini C, Anderson K, Sottili M, Lai M (1991a) Distribution of calcitonin gene-related peptide receptor binding sites (CGRP-RB) in the rat gastrointestinal tract. Gastroenterology 100:A668

Sternini C, De Giorgio R, Furness JB (1991b) Calcitonin gene-related peptide (CGRP) innervation of the canine digestive system. Gastroenterology 100:A668

Stewart-Lee A, Burnstock G (1989) Actions of tachykinins on the rabbit mesenteric artery: substance P and [Glp^6,L-Pro^9]SP$_{6-11}$ are potent agonists for endothelial neurokinin-1 receptors. Br J Pharmacol 97:1218–1224

Stricker S (1876) Untersuchungen über die Gefäßwurzeln des Ischiadicus. Sitzungsber Kaiserl Akad Wiss (Wien) 3:173–185

Su HC, Bishop AE, Power RF, Hamada Y, Polak JM (1987) Dual intrinsic and extrinsic origins of CGRP- and NPY-immunoreactive nerves of rat gut and pancreas. J Neurosci 7: 2674–2687

Sundler F, Ekblad E, Håkanson R (1991) Occurrence and distribution of SP- and CGRP-containing nerve fibers in the gastric mucosa. Species differences. In: Costa M, Surrenti C, Gorini S, Maggi CA, Meli A (eds) Sensory nerves and neuropeptides in gastroenterology. Plenum, New York, pp 29–37

Szolcsányi J (1982) Capsaicin type pungent agents producing pyrexia. In: Milton AS (ed) Pyretics and antipyretics. Springer, Berlin Heidelberg New York, pp 437–478 (Handbook of experimental pharmacology, vol 60)

Szolcsányi J (1983) Disturbance of thermoregulation induced by capsaicin. J Therm Biol 8: 207–212

Szolcsányi J (1984) Capsaicin-sensitive chemoceptive neural system with dual sensory-efferent function. In: Chahl LA, Szolcsányi J, Lembeck F (eds) Antidromic vasodilatation and neurogenic inflammation. Akadémiai Kiadó, Budapest, pp 27–52

Szolcsányi J (1988) Antidromic vasodilatation and neurogenic inflammation. Agents Actions 23:4–11

Szolcsányi J (1990a) Capsaicin, irritation, and desensitization. Neurophysiological basis and future perspectives. In: Green BG, Mason JR, Kare MR (eds) Irritation. Dekker, New York, pp 141–168

Szolcsányi J (1990b) Effect of capsaicin, resiniferatoxin and piperine on ethanol-induced gastric ulcer of the rat. Acta Physiol Hung 75 [Suppl]:267–268

Szolcsányi J, Barthó L (1978) New type of nerve-mediated cholinergic contractions of the guinea-pig small intestine and its selective blockade by capsaicin. Naunyn Schmiedebergs Arch Pharmacol 305:83–90

Szolcsányi J, Barthó L (1981) Impaired defense mechanism to peptic ulcer in the capsaicin-desensitized rat. In: Mózsik G, Hänninen O, Jávor T (eds) Gastrointestinal defense mechanisms. Pergamon, Oxford; Akadémiai Kiadó, Budapest, pp 39–51

Szolcsányi J, Pintér E (1991) Systemic antiinflammatory effect induced by antidromic stimulation of dorsal roots in the rat. Abstracts of the 1st international meeting of the European Neuropeptide Club: functional and pharmacological aspects of neuropeptides, Igls/Innsbruck, p 27

Szolcsányi J, Pintér E, Pethö G (1992) Role of unmyelinated afferents in regulation of microcirculation and its chronic distortion after trauma and damage. In: Jänig W, Schmidt RF (eds) Pathophysiological mechanisms of reflex sympathetic dystrophy. VHC, Mainz, in press

Takaki M, Nakayama S (1989) Effects of capsaicin on mesenteric neurons of the guinea pig ileum. Neurosci Lett 105:125–130

Takeuchi K, Niida H, Okabe S (1990) Gastric motility changes in the cytoprotective action of orally administered N-ethylmaleimide and capsaicin in the rat stomach. Gastroenterology 98:A134

Takeuchi K, Matsumoto J, Ueshima K, Okabe S (1991a) Sensory neuronal desensitization induced duodenal ulcers in the rat in the presence of acid hypersecretion. Gastroenterology 100:A172

Takeuchi K, Matsumoto J, Ueshima K, Okabe S (1991b) Role of capsaicin-sensitive afferent neurons in alkaline secretory response to luminal acid in the rat duodenum. Gastroenterology 101:954–961

Takeuchi K, Niida H, Matsumoto J, Ueshima K, Okabe S (1991c) Gastric motility changes in capsaicin-induced cytoprotection in the rat stomach. Jpn J Pharmacol 55:147–155

Tanaka DT, Grunstein MM (1985) Vasoactive effects of substance P on isolated rabbit pulmonary artery. J Appl Physiol 58:1291–1297

Tepperman BL, Whittle BJR (1992) Endogenous nitric oxide and sensory neuropeptides interact in the modulation of the rat gastric microcirculation. Br J Pharmacol 105:171–175

Terenghi G, Zhang S-Q, Unger WG, Polak JM (1986) Morphological changes of sensory CGRP-immunoreactive and sympathetic nerves in peripheral tissues following chronic denervation. Histochemistry 86:89–95

Theoharides TC, Douglas WW (1978) Somatostatin induces histamine secretion from rat peritoneal mast cells. Endocrinology 102:1637–1640

Thiefin G, Raybould HE, Leung FW, Taché Y, Guth PH (1990) Capsaicin-sensitive afferent fibers contribute to gastric mucosal blood flow response to electrical vagal stimulation. Am J Physiol 259:G1037–G1043

Thom SM, Hughes AD, Goldberg P, Martin G, Schachter M, Sever PS (1987) The actions of calcitonin gene-related peptide and vasoactive intestinal peptide as vasodilators in man in vivo and in vitro. Br J Clin Pharmacol 24:139–144

Tippins JR, Di Marzo VM, Panico M, Morris HR, MacIntyre I (1986) Investigation of the structure-activity relationship of human calcitonin gene-related peptide (CGRP). Biochem Biophys Res Commun 134:1306–1311

Torebjörk HE (1974) Afferent C units responding to mechanical, thermal and chemical stimuli in human nonglabrous skin. Acta Physiol Scand 92:374–390

Törnebrandt K, Nobin A, Owman C (1987) Contractile and dilatory action of neuropeptides on isolated human mesenteric blood vessels. Peptides 8:251–256

Tóth-Kása I, Jancsó G, Obál F, Husz S (1983) Involvement of sensory nerve endings in cold and heat urticaria. J Invest Dermatol 80:34–36

Tóth-Kása I, Katona M, Obál F, Husz S, Jancsó G (1984) Pathological reactions of human skin: involvement of sensory nerves. In: Chahl LA, Szolcsányi J, Lembeck F (eds) Antidromic vasodilatation and neurogenic inflammation. Akadémiai Kiadó, Budapest, pp 317–328

Tóth-Kása I, Jancsó G, Bognár A, Husz S, Obál F (1986) Capsaicin prevents histamine-induced itching. Int J Clin Pharmacol Res 6:163–170

Tramontana M, Cecconi R, Del Bianco E, Santicioli P, Maggi CA, Alessandri M, Geppetti P (1991) Hypertonic media produce Ca^{2+}-dependent release of calcitonin gene-related peptide from capsaicin-sensitive nerve fibres in the rat urinary bladder. Neurosci Lett 124:79–82

Tsutsumi Y, Hara M (1989) Peptidergic nerve fibers in human stomach and duodenal bulb: immunohistochemical demonstration of changes in fiber distribution in chronic gastritis. Biomed Res 10:209–216

Uchida M, Yano S, Watanabe K (1991) The role of capsaicin-sensitive afferent nerves in protective effect of capsaicin against absolute ethanol-induced gastric lesions in rats. Jpn J Pharmacol 55:279–282

Uddman R, Edvinsson L, Ekblad E, Håkanson R, Sundler F (1986) Calcitonin gene-related peptide (CGRP): perivascular distribution and vasodilatory effects. Regul Pept 15:1–23

Umeno E, Nadel JA, McDonald DM (1990) Neurogenic inflammation of the rat trachea – fate of neutrophils that adhere to venules. J Appl Physiol 69:2131–2136

Undem BJ, Myers AC, Weinreich D (1991) Antigen-induced modulation of autonomic and sensory neurons in vitro. Int Arch Allergy Appl Immunol 94:319–324

Uvnäs-Wallensten K (1978) Release of substance P-like immunoreactivity into the antral lumen of cats. Acta Physiol Scand 104:464–468

Van Hees J, Gybels J (1981) C nociceptor activity in human nerve during painful and non painful skin stimulation. J Neurol Neurosurg Psychiatry 44:600–607

Varro A, Green T, Holmes S, Dockray GJ (1988) Calcitonin gene-related peptide in visceral afferent nerve fibres: quantification by radioimmunoassay and determination of axonal transport rates. Neuroscience 26:927–932

Vickers JC, Costa M, Vitadello M, Dahl D, Marotta CA (1990) Neurofilament protein-triplet immunoreactivity in distinct subpopulations of peptide-containing neurons in the guinea-pig coeliac ganglion. Neuroscience 39:743–759

Walder CE, Thiemermann C, Vane JR (1990) Endothelium-derived relaxing factor participates in the increased blood flow in response to pentagastrin in the rat stomach mucosa. Proc R Soc Lond [Biol] 241:195–200

Wallace JL (1990) 5-Lipoxygenase: a rational target for therapy of inflammatory bowel disease? Trends Pharmacol Sci 11:51–53

Wallengren J (1991) Substance P antagonist inhibits immediate and delayed type cutaneous hypersensitivity reactions. Br J Dermatol 124:324–328

Wallengren J, Håkanson R (1987) Effects of substance P, neurokinin A and calcitonin gene-related peptide in human skin and their involvement in sensory nerve-mediated responses. Eur J Pharmacol 143:267–273

Wallengren J, Möller H (1988) Some neuropeptides as modulators of experimental contact allergy. Contact Dermatitis 13:351–354

Wallengren J, Ekman R, Möller H (1986) Substance P and vasoactive intestinal peptide in bullous and inflammatory skin disease. Acta Derm Venereol (Stockh) 66:23–28

Wallengren J, Ekman R, Sundler F (1987) Occurrence and distribution of neuropeptides in the human skin. An immunochemical and immunocytochemical study on normal skin and blister fluid from inflamed skin. Acta Derm Venereol (Stockh) 67:185–192

Walmsley D, Wiles PG (1991) Early loss of neurogenic inflammation in the human diabetic foot. Clin Sci 80:605–610

Wanaka A, Matsuyama T, Yoneda S, Kimura K, Kamada T, Girgis S, MacIntyre I, Emson PC, Tohyama M (1986) Origins and distribution of calcitonin gene-related peptide-containing nerves in the wall of the cerebral arteries of the guinea-pig with special reference to the coexistence with substance P. Brain Res 369:185–192

Weihe E (1990) Neuropeptides in primary afferent neurons. In: Zenker W, Neuhuber WL (eds) The primary afferent neuron. Plenum, New York, pp 127–159

Wharton J, Gulbenkian S, Mulderry PK, Ghatei MA, McGregor GP, Bloom SR, Polak JM (1986) Capsaicin induces a depletion of calcitonin gene-related peptide (CGRP)-immunoreactive nerves in the cardiovascular system of the guinea-pig and rat. J Auton Nerv Syst 16:289–309

White DM, Helme RD (1985) Release of substance P from peripheral nerve terminals following electrical stimulation of the sciatic nerve. Brain Res 336:27–31

Whittle BJR (1977) Mechanisms underlying gastric mucosal damage induced by indomethacin and bile salts, and the actions of prostaglandins. Br J Pharmacol 60:455–460

Whittle BJR, Lopez-Belmonte J (1991) Interactions between the vascular peptide endothelin-1 and sensory neuropeptides in gastric mucosal injury. Br J Pharmacol 102: 950–954

Whittle BJR, Lopez-Belmonte J, Rees DD (1989) Modulation of the vasodepressor actions of acetylcholine, bradykinin, substance P and endothelin in the rat by a specific inhibitor of nitric oxide formation. Br J Pharmacol 98:646–652

Whittle BJR, Lopez-Belmonte J, Moncada S (1990) Regulation of gastric mucosal integrity by endogenous nitric oxide: interactions with prostanoids and sensory neuropeptides in the rat. Br J Pharmacol 99:607–611

Williams G, Cardoso H, Ball JA, Mulderry PK, Cooke E, Bloom SR (1988) Potent and comparable vasodilator actions of A- and B-calcitonin gene-related peptides on the superficial subcutaneous vasculature of man. Clin Sci 75:309–313

Williams TJ (1982) Vasoactive intestinal polypeptide is more potent than prostaglandin E_2 as a vasodilator and oedema potentiator in rabbit skin. Br J Pharmacol 77:505–509

Williamson TE, Donaldson CL, Morris GP (1990) "Referred" adaptive cytoprotection is prostaglandin-dependent but "local" cytoprotection is not. Gastroenterology 98:A149

Wimalawansa SJ, MacIntyre I (1988) Calcitonin gene-related peptide and its specific binding sites in the cardiovascular system. Int J Cardiol 20:29–37

Woodward DF, Owen DA, Pipkin MA, Ledgard SE (1985) Histamine involvement in the local and systemic microvascular effects produced by intradermal substance P. Agents Actions 17:126–130

Wormser U, Laufer R, Hart Y, Chorev M, Gilon C, Selinger Z (1986) Highly selective agonists for substance P receptor subtypes. EMBO J 5:2805–2808

Xu X-J, Hao J-X, Wiesenfeld-Hallin Z, Håkanson R, Folkers K, Hökfelt T (1991) Spantide II, a novel tachykinin antagonist, and galanin inhibit plasma extravasation induced by antidromic C-fiber stimulation in rat hindpaw. Neuroscience 42:731–737

Xu X-J, Dalsgaard C-J, Maggi CA, Wiesenfeld-Hallin Z (1992) NK-1, but not NK-2, tachykinin receptors mediate plasma extravasation induced by antidromic C-fiber stimulation in rat hindpaw: demonstrated with the NK-1 antagonist CP-96,345 and the NK-2 antagonist Men 10207. Neurosci Lett 139:249–252

Yaksh TL (1988) Substance P release from knee joint afferent terminals: modulation by opioids. Brain Res 458:319–324

Yano H, Wershil BK, Arizono N, Galli SJ (1989) Substance P-induced augmentation of cutaneous vascular permeability and granulocyte infiltration in mice is mast cell dependent. J Clin Invest 84:1276–1286

Yeo CJ, Jaffe BM, Zinner MJ (1982a) Local regulation of blood flow in the feline jejunum. A possible role for endoluminally released substance P. J Clin Invest 70:1329–1333

Yeo CJ, Zinner JM, Denoy D, Jaffe BM (1982b) Intravenous substance P infusion in conscious dogs. Its effect on hemodynamics, regional blood flow, glucose, insulin, and cortisol. Gastroenterology 82:1216

Yokosawa H, Endo S, Ogura Y, Ishii S (1983) A new feature of angiotensin converting enzyme in the brain: hydrolysis of substance P. Biochem Biophys Res Commun 116: 735–742

Yokotani K, Fujiwara M (1985) Effects of substance P on cholinergically stimulated gastric acid secretion and mucosal blood flow in rats. J Pharmacol Exp Ther 232:826–830

Yonehara N, Shibutani T, Inoki R (1987) Contribution of substance P to heat-induced edema in rat paw. J Pharmacol Exp Ther 242:1071–1076

Yonehara N, Imai Y, Inoki R (1988) Effect of opioids on the heat stimulus-evoked substance P release and thermal edema in the rat hind paw. Eur J Pharmacol 151: 381–388

Yonehara N, Imai Y, Inoki R (1991) Influence of heat stimulation on the amount of calcitonin gene-related peptide and neurokinin A in the subcutaneous space of the rat hind instep. Jpn J Pharmacol 56:381–384

Zaidi M, Bevis PJR, Girgis SI, Lynch C, Stevenson JC, MacIntyre I (1985) Circulating CGRP comes from the perivascular nerves. Eur J Pharmacol 117:283–284

Zawadzki JV, Furchgott RF, Cherry PD (1981) The obligatory role of endothelial cells in the relaxation of arterial smooth muscle by substance P. Fed Proc 40:689

Ziche M, Morbidelli L, Pacini M, Geppetti P, Alessandri G, Maggi CA (1990) Substance P stimulates neovascularization in vivo and proliferation of cultured endothelial cells. Microvasc Res 40:264–278

Zimmerman BJ, Gaginella TS, Granger DN (1991) Substance P: a modulator of leukocyte adhesion in rat mesenteric venules. Gastroenterology 100:A265

Zimmerman RP, Gates TS, Mantyh CR, Vigna SR, Welton ML, Passaro EP, Mantyh PW (1989) Vasoactive intestinal polypeptide receptor binding sites in the human gastrointestinal tract: localization by autoradiography. Neuroscience 31:771–783

Zimmermann M (1979) Peripheral and central nervous mechanisms of nociception, pain, and pain therapy: facts and hypotheses. In: Bonica JJ, Liebeskind JC, Albe-Fessard DG, Jones LE (eds) Advances in pain research and therapy, vol 3. Raven, New York, pp 3–32

Rev. Physiol. Biochem. Pharmacol., Vol. 121
© Springer-Verlag 1992

Molecular Biology of the Human Y Chromosome

ULRICH WOLF, WERNER SCHEMPP, and GERD SCHERER

Contents

Institut für Humangenetik und Anthropologie der Universität, Breisacherstraße 33,
W-7800 Freiburg i.Br., FRG

We dedicate this review to Susumu Ohno, whose ideas motivated our interest in this exciting field of research

1 Introduction

The Y chromosome is involved in sex determination and male sexual differentiation in human beings as in other mammalian species. Presence of this chromosome normally results in male development, while in its absence a female phenotype develops. There are various conditions of sex reversal and intersexuality which are due to mutations of all possible kinds at the chromosome and gene level. Their analysis has finally resulted in the definition of a gene on the Y chromosome fulfilling all the requirements for a major testis-determining gene, which codes for the long sought-after testis-determining factor (TDF).

While the term "sex determination" is used in this article for a process which triggers off a cascade of events resulting in sex-specific gonadal differentiation, the term gonad differentiation refers to gonadal morphogenesis as the consequence of sex determination.

The role of the Y chromosome in sex determination and male sexual differentiation is the obvious reason for the specialization of the Y chromosome during evolution. It belongs to the smallest elements of the chromosome complement, but it is still not specialized as to transport only determinants for male development. Due to its origin and evolutionary fate, the human Y chromosome includes numerous other sequences which are not unique to this chromosome. To understand its present organization, it is therefore necessary to compare it with the X chromosome and to trace its evolution by considering other species possessing Y chromosomes which may reflect ancestral situations.

The molecular biology of the Y chromosome has been the subject of several books and reviews over the past years, among which the article by Goodfellow et al. (1985a) and the books edited by Sandberg (1985), Goodfellow et al. (1987a), McLaren and Ferguson-Smith (1988), and Wachtel (1989) may be mentioned here. In addition, mapping data are regularly compiled in the reports of the conferences on Human Gene Mapping (HGM), which have taken place from 1973 onwards, and are published in *Cytogenetics and Cell Genetics* as separate issues. In the first two conference reports (HGM1 1974; HGM2 1975), there is no assignment of a gene to the Y chromosome. In HGM3 (1976) the only gene tentatively assigned to the Y chromosome is that for H-Y antigen. The number of markers increased slowly, including protein markers, DNA sequences, and phenotypic characters, and up to HGM9 (1987), the Y chromosome was not covered separately from the X chromosome. Only from HGM10 (1989) onwards was a special committee on the genetic constitution of the Y chromosome appointed, and the last report of this committee (HGM11 1991) refers already to some 230 different DNA segments, 12 genes or pseudogenes, and three phenotypic characters. Of the approximately 230 DNA segments, 121 are Y-specific single-copy loci, 35 of which were assigned to the short arm and 83 to the long arm, while three are of unknown location.

Though we have learned much about the genetic organization and function of the Y chromosome, in particular by the very powerful combination of molecular and cytogenetic methods, we are still far from a comprehensive understanding of the mechanism of sex determination and of the genetic significance of large parts of this chromosome. Since current research is rather rapidly elucidating these issues, only an instantaneous picture of present knowledge can be drawn, which will already be outdated while this article is being written.

2 Origin of the Y Chromosome

In evolutionary terms, we consider a Y chromosome "primitive" if it is still largely homologous to the X chromosome, pointing to its origin from a homologous pair of autosomes. The most advanced Y chromosome would be a tiny element specialized largely or exclusively in male sexual development. The human Y chromosome does not fit into either scheme. It is not "primitive" because ancestral linkage groups reflecting its autosomal origin are scarcely conserved, but it is also not specialized for carrying only male-determining genes because it has acquired secondarily numerous sequences of various origins. Therefore, it represents a modern chromosome, and it reached its present configuration during the divergent evolution of the branches leading to the great apes and to man. By studying primate Y chromosomes, we can try to obtain a picture of the evolutionary rearrangements which have taken place, resulting in the present-day human Y chromosome. However, in the entire infraclass eutheria (placental mammals) the Y chromosome, apart from its specialization for male development, appears to have undergone varying secondary changes, and to trace its evolution we must therefore look beyond this taxonomic border. While the infraclass metatheria (marsupials) is not very informative in this respect, showing a tiny Y chromosome which appears to be far more specialized in sex determination than the eutherian Y, study of the subclass prototheria (monotremes) is most rewarding. This group possesses a rather primitive Y chromosome which appears to be still largely, but not entirely, homologous to the X chromosome. This homology, if confirmed, is a key finding for an understanding of mammalian sex chromosome evolution, pointing to the origin of the sex chromosomes from a homologous pair of autosomes. While the assumption of X-Y homology in prototheria is based on recent observations, chromosome studies in various groups of nonmammalian vertebrates prompted Ohno, as early as 1967, to postulate that sex chromosome evolution started with an autosomal chromosome pair.

It appears that during vertebrate evolution, sex chromosome differentation, resulting in a largely conserved X (Z) chromosome and a more or less specialized Y (W) chromosome characteristic for the heterogametic sex, has taken place several times independently, and this process may have started with a different pair of autosomes in each case. (If the heterogametic sex is male, XY is used, if it is female, ZW, to designate the sex chromosomes.) We can assume that the sex chromosome heteromorphism of mammals originated during the evolution of this vertebrate class itself, and so we cannot expect to find ancestral stages of this process among nonmammalian vertebrates. However, similar processes which have taken place in other taxonomic groups of vertebrates may well serve as models for mammals. Particularly illuminating are some examples from amphibians and reptiles which will briefly be described below. Because of the presumed ancestral homology between the sex chromosomes, Y chromosome evolution cannot be considered on its own; its relation to the X chromosome has to be included for comparison.

2.1 Evolution of the Sex Chromosomes

2.1.1 Nonmammalian Vertebrates as Models

A model of sex chromosome evolution, which can now be considered classical, was provided by Beçak et al. (1964) and further elaborated by Jones, Singh, and associates in a series of beautiful studies on the repitilian suborder Ophidia (snakes), exemplifying many states in a series of gradual

changes from an undifferentiated homomorphic pair to a markedly heteromorphic pair of chromosomes (for review see Jones and Singh 1985).
The mechanisms of sex determination in snakes is of the ZW type, i.e., the
male is the homogametic (ZZ) and the female the heterogametic (ZW) sex.
In species of the most ancient and primitive family Boidae, sex chromosomes cannot be identified; they must be a homomorphic and largely autosomal pair of chromosomes. In the family Colubridae, representing more
advanced snakes, some species still are endowed with homomorphic sex
chromosomes, others show a pericentric inversion in their W chromosome,
and still others have a significantly smaller W chromosome as compared
with the Z (reviewed in Ohno 1967). Interestingly, in all species with Z/W
heteromorphism, the W is condensed in the interphase nucleus, forming a
chromocenter, and it shows asynchronous late replication behavior (Ray-
Chaudhuri et al. 1970, 1971). These characteristics are consistent with
functional inactivation of the W. The analysis of the W chromosome was
further advanced by the finding, using in situ hybridization techniques, that
highly repetitive satellite DNA was concentrated on the W chromosome
depending on its state of differentiation. Thus, in primitive Boidae, the hybridization pattern revealed no sex difference. However, among colubrid
snakes, even in species with homomorphic sex chromosomes, the W showed significant hybridization with the satellite DNA (or complementary
RNA) and was identified in this way (Singh et al. 1976, 1980). From these
findings it was concluded that the occurrence of satellite DNA is the first
step in the divergent evolution of the W from the Z chromosome, resulting
eventually in asynchronous DNA replication and meiotic isolation. Indeed,
it has been shown in oocytes of a species with heteromorphic sex chromosomes that the W is condensed during meiotic prophase, preventing Z and
W chromosomes from crossing over (Singh et al. 1979). This meiotic isolation of the W chromosome was no doubt the prerequisite for its subsequent
reduction in size and specialization for the determination of the heterogametic sex.

The hypothesis, derived from the study of snakes, that a change in
molecular constitution initiated the specialization process of the chromosome determining the heterogametic sex is further illustrated by studies of
the DNA replication patterns of the frog species *Rana esculenta* (Schempp
and Schmid 1981). While no sex chromosomes were identified by inspec-
tion of chromosome structure, BrdU replication patterns revealed a late-replicating segment in one member of chromosome pair number 4 in the
male only. This asynchrony shows that the male is the heterogametic sex,
and that the late-replicating chromosome 4 functions as a Y chromosome.
Similar examples have been reported from other taxonomic groups, e.g., li-
zards (Olmo et al. 1984). It can be concluded that functional changes (re-

plication asynchrony, heteropyknosis, inactivation) resulting in meiotic isolation preceded structural changes in the evolution of sex chromosome heteromorphism.

2.1.2 Mammals

The examples chosen in the preceding section may serve as models for a similar process during mammalian sex chromosome evolution. However, the mechanisms may vary in detail, as is the case with these examples. In snakes, highly repetitive sequences first invaded the chromosome determining the heterogametic sex, then became spread and amplified over its entire length, and may thus be the reason for its functional change, whereas in *Rana esculenta* (as in other species) a shift in the DNA replication period may be the first step towards functional change. (It must be mentioned, however, that in these latter cases satellite DNAs have not been studied.) Therefore, in mammals one of these or still another mechanism may have been operating. In analyzing the mammalian system, we depend on the rather limited variation occuring between contemporaneous species. It is not to be expected that more primitive stages can be found among nonmammalian vertebrates because, as discussed above, the evolutionary process resulting in the sex chromosomes of recent mammals must have taken place during premammalian evolution itself.

Mammals are divided into two subclasses, prototheria (monotremata) and theria, and the latter subclass into the infraclasses metatheria (marsupialia) and eutheria (placentalia). A chromosomal mechanism of sex determination operates throughout, and it is of the XX/XY type. While there are extensive homologies between the X chromosomes of all three subclasses, the Y chromosomes differ with respect to their specialization in determining the male sex. It appears that the prototheria still possess a rather primitive Y of considerable size. In one species (platypus) it is nearly the same length as the X, and it is homologous to the X except for the short arm by cytogenetic criteria (Wrigley and Grages 1988). In eutheria, the Y is considerably reduced in size compared with the X, but it is still composed of various segments showing sequence homology to the X and to autosomes. Thus, it represents an intermediate type of specialization. The most advanced Y chromosomes are found in metatheria; these chromosomes show no homology with other chromosomes at the cytogenetic level and are relatively minute elements (Rofe and Hayman 1985). The present view on the evolution of the mammalian sex chromosomes is depicted in Fig. 1.

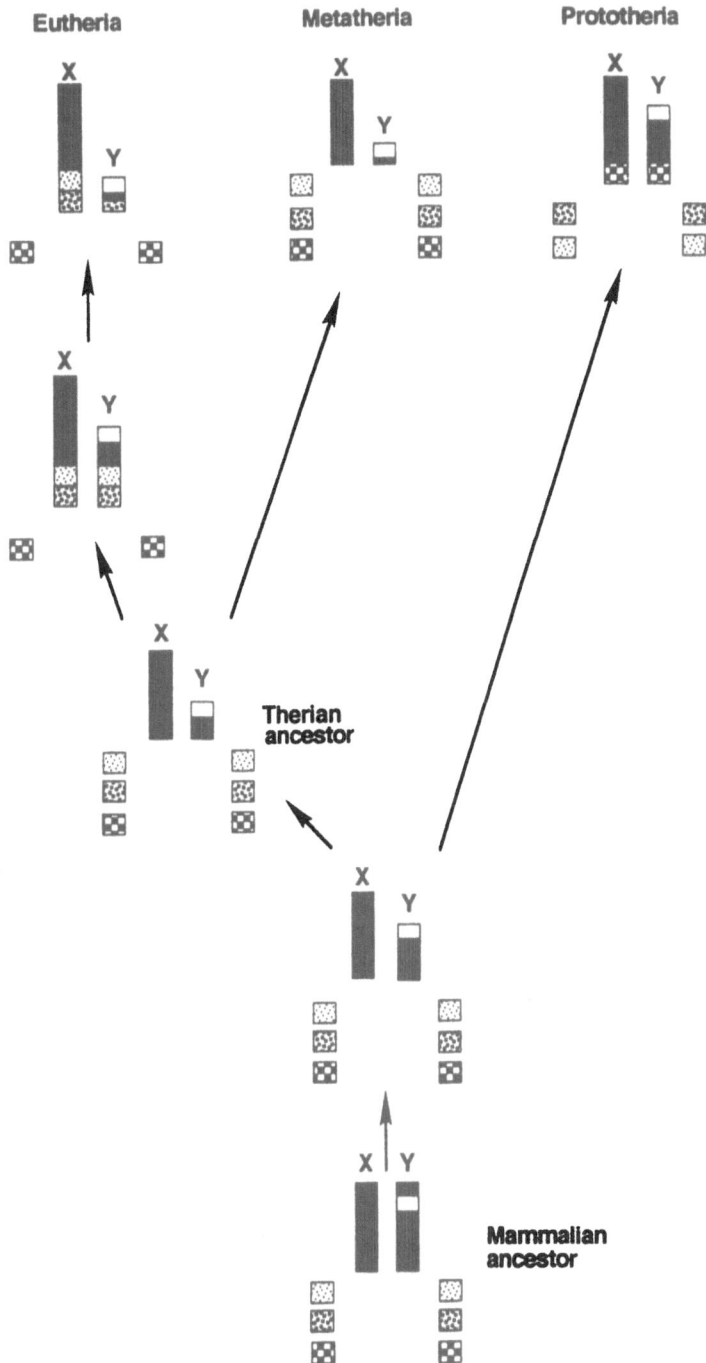

Fig. 1. Hypothetical evolution of mammalian sex chromosomes from an ancestral homo-morphic chromosome pair. The original chromosomes are homologous (*solid black*) with the exception of a linkage group involved in male sexual determination and differentiation (*white block*). The prospective X chromosome, comprising 3% of the haploid female com-

plement, was conserved throughout evolution. The prospective Y chromosome was largely conserved in prototherians, while in the therian infraclasses only the male-specific linkage group was conserved (*white block*). Three independent linkage groups are assumed to have been autosomal in the mammalian ancestor (*stippled squares* and *checkerboard square*). One of these autosomal blocks was transposed to the sex chromosomes in prototherians (*checkerboard*), the two others remaining autosomal. In metatherian evolution, all three blocks remained autosomal. On the way to eutherians, two of these autosomal blocks (*stippled squares*) are assumed to have been transposed to the sex chromosomes and are found on the extant eutherian X chromosome, while they have been largely lost from the Y chromosome. The complex reorganization of the Y chromosome during eutherian evolution is not considered here. (Modified from Graves and Watson 1991)

It would be most straightforward to conclude that these extant Y chromosomes represent the stages to be expected if their evolution started with a homologous chromosome pair, the X chromosome remaining conserved, and the Y chromosome becoming increasingly specialized for male determination. This view has at least to be qualified in various details. It must be recognized that since all three systems have an equally long time span in which to evolve, completely different interpretations are possible. To discuss this issue, it may be useful to begin with the problem of the conservation of the X chromosome, as this should help to define the ancestral situation.

It has long been noted that in eutheria, the X/autosome ratio is nearly constant, with the X chromosome comprising some 5% of the haploid female genome (Ohno et al. 1964). Comparative mapping of X-linked genes has revealed extensive genetic conservation of this chromosome in all eutherian mammals studied, and no exception has been found so far (O'Brien and Graves 1991). Within metatheria, the X chromosome is also conserved, but it comprises only some 3% of the haploid female genome (Hayman and Martin 1974). Comparative mapping has shown that a large linkage group of the eutherian X chromosome is also conserved in the metatherian X (Watson et al. 1990). In terms of the human X chromosome, this linkage group is located on the long arm (Xq), which again comprises 3% of the genome. Thus, it appears that the metatherian X represents the long arm of the human X chromosome. Genes mapped on the short arm of the human X chromosome, however, are clustered in two autosomal blocks in metatheria (Spencer et al. 1991; Watson et al. 1991). Turning to prototheria, their X chromosome is the largest among all mammals, comprising 6% of the genome (Watson 1990). The available mapping data again show the linkage group of the human Xq conserved on the prototherian X, and the loci homologous to the human Xp genes are concentrated in two clusters on autosomes, similar to metatheria (Watson et al. 1991). The larger size of the prototherian X may represent the original state of the ancestral chromosome, or autosomal segments may have been translocated onto it subsequently. A review of these evolutionary relationships was recently provided by Graves and Watson (1991) (Fig. 1).

From these data, it is to be concluded that a linkage group of some 3% of the genome was conserved in all mammals, corresponding to the long arm of the human X chromosome. It seems quite likely that this segment reflects an original part of the ancestral homologous pair of autosomes which differentiated into the mammalian sex chromosomes during evolution. If so, it is of extreme interest to know whether genes belonging to this presumptive ancestral linkage group can be mapped on the Y chromosome. The tiny metatherian Y is no candidate for such investigations, and the largely reduced eutherian Y has undergone massive secondary changes, as will be discussed below. However, the prototherian Y is a large chromosome comprising about 4% of the haploid genome, and the analysis of chromosomal banding patterns indicates that its long arm is homologous to that of the X chromosome long arm (Graves and Watson 1991). In addition, while the short arms of the prototherian sex chromosomes replicate their DNA asynchronously and do not pair at meiosis, the long arms show synchronous replication and pair at meiosis, which is strong evidence for genetic homology. However, information on the genetic content of the prototherian Y chromosome is still lacking.

2.2 The Problem of the Conservation of Pre-eutherian Genes

Regarding, the situation in prototheria, X-Y homology appears still to be preserved to a larger extent, supporting the hypothesis that the sex chromosomes originate from a homologous pair of autosomes. The eutherian Y chromosome, though one of the smallest elements of the karyotype, still comprises about 2% of the haploid complement, containing about 60 megabases (Mb) of DNA (Morton 1991). Thus, there is ample space left for genes other than those responsible for male sexual development, and the question arises whether these genes or some of them represent an ancestral linkage group which was conserved on the Y chromosome and is homologous to a segment of the long arm of the X chromosome.

Of the 12 Y-linked genes and pseudogenes identified so far in man (see Sect. 5 below), seven have a homologue on the X chromosome, six of which map to the short arm (Xp). Of these six genes, five (CSF2RA, MIC2, STS, ZFX, and KALIG-1X/ADMLX) escape inactivation, and for the sixth (AMGX) indications for partial escape from inactivation exist (see Table 1, p. 165). It can be assumed that these genes on Xp do not represent remainders of the autosomal origin of the sex chromosome pair, but rather are later additions to the sex chromosomes, and their escape from inactivation may reflect their more recent autosomal nature. This view is supported by the finding, already mentioned, that genes on the short arm of the human X chromosome are clustered in two autosomal blocks in meta- and prototheria (Graves and Watson 1991).

Recently, a gene coding for a ribosomal protein and named RPS4 was identified on the Y short arm. It has a homologue on the X long arm, located on Xq13 in the same region as the gene for phosphoglycerate kinase (PGK1), and close to the gene for the androgen receptor (AR) on Xq12 (Fisher et al. 1990b). Both genes, PGK1 and AR have been shown to belong to the ancestral Xq linkage group of proto- and metatheria (O'Brien and Graves 1991), while for RPS4 this is known only in metatheria (J.A.M. Graves, personal communication). However, there is a remarkable difference between these genes with respect to their inactivation behavior. While AR and PGK1 are subject to X-inactivation, RPS4 escapes inactivation. Therefore, RPS4 does not appear to be a member of this old linkage group and may have been transposed onto the sex chromosomes during early mammalian evolution. It cannot be excluded, however, that RPS4 is indeed an original gene on the X and Y chromosomes, thus representing the only gene known so far on the human Y chromosome, apart from male-determining genes, which was conserved throughout the evolution of the Y chromosome.

There are a number of anonymous DNA sequences sharing X-Y homology, including those which map to the X chromosome long arm. Some of the latter have been shown to have homologies on the X chromosome of various primates, but in these species those sequences are absent from the Y chromosome. Therefore, they must have been translocated very recently onto the Y during human evolution (see Sect. 3 below).

Thus, regarding the ancestral linkage group of mammalian sex chromosomes, it can be stated that so far no DNA segment of the eutherian Y chromosome has been definitely identified which still reflects the autosomal origin of this chromosome. It may well be the case that the eutherian Y chromosome passed through a stage in evolution where it represented a minute chromosome specialized in male development, similar to that in extant metatheria, and that its present state is the result of secondary rearrangements, including translocations derived from the X chromosome and from autosomes. However, it should have included a pairing segment of effective size allowing for meiotic recombination with the X chromosome as a prerequisite for a translocation homologous to both X and Y chromosomes (Fig. 1). If so, the process of specialization would have taken place after the theria diverged from prototheria, and the metatherian Y chromosome would represent a fragment of the ancestral therian Y only, while the extant eutherian Y is a composite chromosome due to secondary rearrangements. The only ancestral linkage group of the eutherian Y chromosome would then comprise genes involved in male sexual development, including TDF, and one or several genes necessary for male germ cell differentiation ("fertility factors"), like the gene for germ cell motility (AZF) which was assigned to the human Y chromosome. It would be of interest to know if, in the metatherian Y chromosome, in addition to TDF such fertility genes are present.

2.3 Rearrangements in Primates

Following the argumentation of the foregoing section, evolution of the eutherian Y chromosome started with a considerably reduced element containing genes for male differentiation and, presumably, a segment homologous to the X chromosome allowing for recombination. However, we know the eutherian Y as an element of considerable size, comprising about 2% of the haploid genome; thus, the question arises whether the origin of these secondarily acquired DNA segments can be traced. We confine ourselves to primates because the human descendence beyond primates is heavily debated, and it would be arbitrary to refer to extant species of the orders Insectivora, Edentata, or Scandentia, which are considered to be closest to the evolutionary branch leading to primates. In addition, mapping data on the Y chromosomes of these taxonomic groups are not available.

As landmarks, the pseudoautosomal region (PAR) and the sex-determining region (SDR) may be used first to trace rearrangements of the Y chromosome in primates. In man both regions are located in close proximity to each other on the terminal short arm of the Y chromosome (Vergnaud et al. 1986; Sinclair et al. 1990).

The ZFY gene is in close proximity to the SDR on the human Y; assuming that it can thus serve as a marker for this region, Müller and Schempp (1991) have shown that, as in the human being, the ZFY gene maps close to the early-replicating pseudoautosomal segment in telomeric position of the Y chromosomes of the great apes. Thus, despite cytogenetically visible structural alterations within the euchromatic parts of the Y chromosomes of man and the great apes (Weber et al. 1986, 1987), the close proximity of the PAR and, by inference, of the SDR is conserved throughout the human and great-ape lineages. This situation contrasts to that in the mouse, in which species these segments are located on opposite sides of the Y chromosome, the SDR being on the short arm (Gubbay et al. 1990) and the PAR representing the distal long arm segment (Keitges et al. 1985; Harbers et al. 1986).

Recent molecular analysis of the steroid sulfatase (STS) locus may explain some of the human-mouse differences. In human beings the X-linked STS gene is located about 5 Mb from the PAR and escapes X-inactivation, while a nonfunctional pseudogene of STS is located on the long arm of the Y chromosome (Yen et al. 1988; see Sect. 5.9). Interestingly, in the mouse Sts is pseudoautosomal (Keitges et al. 1985). Comparative studies in primates strongly suggest that the ancestral STS gene was pseudoautosomal and that a pericentric inversion in the early primate lineage disrupted the former pseudoautosomal location of STS (Yen et al. 1988; Schempp and Toder 1992). An analogous situation was demonstrated for the Kallmann (KALIG-1) locus (see Sect. 5.8). In human beings the X-linked KALIG-1 gene is located some 1.5 Mb proximal to the STS gene and likewise escapes X-inactivation, while a homologue on the long arm of the Y chromosome presumably is not expressed (Franco et al. 1991; Legouis et al. 1991). Again, comparative studies in primates indicate that the ancestral KALIG-1 gene was pseudoautosomal and became sex specific subsequent to a pericentric inversion in the Y chromosome (Franco et al. 1991).

These data strongly suggest that a pericentric inversion occurred after the divergence of the prosimian and the simian lineages which disrupted the ancestral PAR, transferring its proximal part, including STS and KALIG-1, to the long arm of the Y chromosome. Simultaneously, this pericentric inversion brought the SDR from the long arm into close proximity to the PAR on the short arm of the Y chromosome (Fig. 2).

Further cytogenetically visible structural rearrangements of the Y chromosome of various simian species have been described, locating an early replicating segment (the presumed pairing segment with the X chromosome in meiosis) on the telomeric long or short arm of the Y chromosome (Weber et al. 1987; Schempp et al. 1989b). These rearrangements seem to characterize secondary events which have occurred in addition to the above-described "primary" pericentric inversion of the Y chromosome.

There are other rearrangements which must have occurred during the evolution of the human Y chromosome, and it can be stated that, while the eutherian X chromosome was conserved in its entirety, the Y chromosome was subject to enormous variation not only in structure, but also with respect to its DNA composition. The various DNA components of the human Y will now be discussed, and in this connection further comparisons with other mammalian species will be of some value.

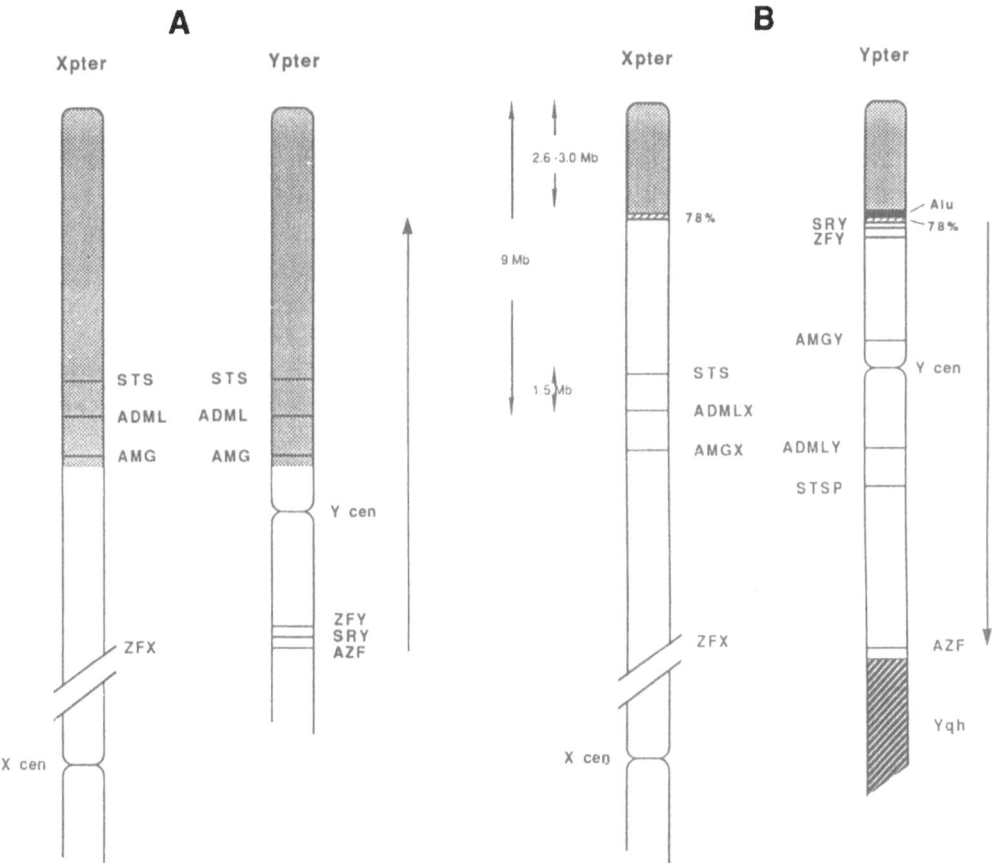

Fig. 2 A, B. A pericentric inversion in the Y chromosome disrupts an originally contiguous large pseudoautosomal region and an ancestral male-specific linkage group. **A** Ancestral situation: The pseudoautosomal segment (*stippled*) includes the present-day pseudoautosomal region and an adjacent segment represented by the STS, ADML, and AMG genes which also behaved pseudoautosomally. The sex-specific region is on the opposite side of the Y centromere. **B** Present-day situation: By a pericentric inversion in the Y chromosome, the proximal part of the pseudoautosomal region including the STS, ADML, and AMG genes was transposed to the long arm, while part of the sex-specific region came into close proximity to the distal part of the pseudoautosomal region. The location of AMGY on proximal Yp is assumed to be due to a secondary pericentric inversion. Genes on the displaced pseudoautosomal segment no longer behave pseudoautosomally. The present-day pseudoautosomal boundary on the Y chromosome is marked by an Alu element and a proximal segment of 78% sequence identity between X and Y, at the proximal end of which sequence similarity ceases abruptly. Physical distances between some loci are indicated along the distal X chromosome short arm

3 X-Y Homologies

During the meiotic prophase, the human X and Y chromosomes are not paired along their entire length like the autosomes; pairing and formation of a synaptonemal complex are usually restricted to the short arms, starting at the telomeric ends. The pairing region is variable and may extend even into the long arm of the Y chromosome (Chandley et al. 1984), but it regularly includes the distal part of the short arm. Indeed, crossing-over is normally observed only in that distal part (Hultén 1974, Solari 1980), even if the synaptonemal complex also comprises proximal segments. Based on these observations, it was concluded that genetic homology exists only for the segment undergoing crossing-over, while pairing may also include nonhomologous parts of the sex chromosomes. Moreover, indications for a functional homology of the distal short arms of X and Y also come from DNA replication studies using the bromodeoxyuridine-incorporation technique; these segments show synchronous early replication during the mitotic S-phase (Schempp and Meer 1983). Thus, cytogenetically, a homologous segment comprising the distal short arm and a differential segment comprising all the rest of the Y chromosome can be identified. From DNA analysis, we now know that the distal short arms of X and Y, starting from the telomeres, indeed represent homologous colinear segments showing genetic recombination. Thus, this segment behaves like an autosomal segment, and the term "pseudoautosomal region", originally proposed by Burgoyne (1982) has therefore found wide acceptance. However, other genes distributed along the Y chromosome have a homologous counterpart on the X chromosome, and evidently there is at least one larger segment on the proximal long arm of the Y recombining occasionally with a homologous segment on Xp, which may represent a region of pseudoautosomal origin.

3.1 The Pseudoautosomal Region

The assumption of a homologous region at the distal short arms of the X and Y chromosomes was originally derived from cytogenetic observations in the rat, i.e., pairing and crossing-over at meiosis (Koller and Darlington 1934). This assumption was confirmed in the human species when genetic markers were assigned to this region, allowing for testing of recombination. With the increasing number of those markers, comprising in the vast majority anonymous DNA segments, the pseudoautosomal region was more closely defined, including its extension, the nature of its boundary to the differential segment, and recombination fractions of various markers with respect to the sexual phenotype. In addition, some insights into the origin of the pseudoautosomal region were obtained (for review, see Ellis and Goodfellow 1989).

According to the data presented at the last conference on human gene mapping (Weissenbach and Goodfellow 1991) two genes, 13 single copy DNA segments, and two multiple copy DNA segments were definitely assigned to the pseudoautosomal region. In addition, a large number of DNA probes detecting loci of X-Y homology which were assigned to the short arm of the Y chromosome, but not closer localized, may well turn out to map in the pseudoautosomal region.

The first genetic evidence for sequence homology between the sex chromosomes was provided by the study of a gene encoding a cell surface antigen of unknown function, designated MIC2, which was detected by a monoclonal antibody raised against cells of a human T-cell leukemia (Goodfellow et al. 1983). MIC2 was subsequently shown by in situ hybridization to be located in the pairing region at the distal ends of Xp and Yp (Buckle et al. 1985), to be polymorphic, and to recombine with sex (locus for TDF) at a frequency between 2.5% (Goodfellow et al. 1986) and 0% (Johnson et al. 1991). Thus, MIC2 fulfills the requirements for a pseudoautosomal gene. However, the recombination frequency is far from independent, and the strong sex linkage of the MIC2 gene reflects its location near the boundary of the pseudoautosomal region and the differential part of the Y chromosome.

Indeed, the study of other polymorphic DNA sequences from the pseudoautosomal region has revealed a gradient of sex linkage (Simmler et al. 1985; Rouyer et al. 1986), the most distal marker – the telomeric locus DXYS14 – showing random segregation with sex (Cooke et al. 1985), while other polymorphic markers are partially sex linked, with decreasing recombination frequencies the more proximal their location (Fig. 3 B).

In addition to MIC2, another gene has recently been localized within the pseudoautosomal region, CSF2RA, which stands for "colony-stimulating factor 2 receptor alpha," or granulocyte-macrophage colony-stimulating factor receptor (Gough et al. 1990). This gene is polymorphic, showing partial sex linkage. However, in contrast to MIC2, with only 0%–2.5% recombination, it recombines with a higher frequency of about 20%. Correspondingly, CSF2RA is located more distally within the pseudoautosomal region, between the loci DXYS15 and DXYS17, about 1.2–1.3 Mb from the telomere (Rappold et al. 1991; see Fig. 3 B).

The free recombination of the telomeric marker DXYS14 (Cooke et al. 1985) confirms the prediction of Koller and Darlington (1934), stressed more recently by Burgoyne (1982), of an obligatory chiasma within the X/Y pairing region. In this connection, it should be mentioned that in man,

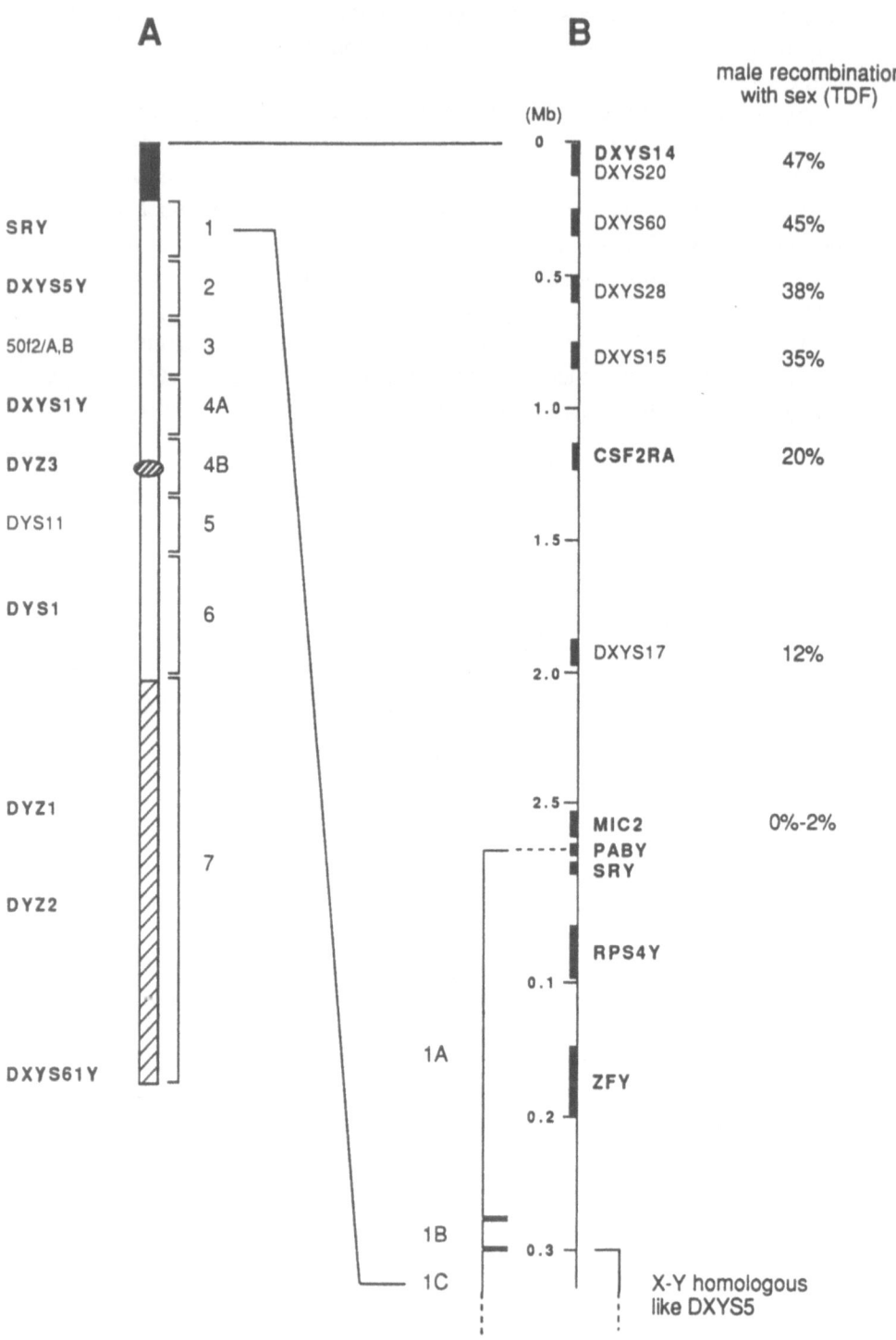

Fig. 3 A, B. Schematic interval map of the Y chromosome and physical map of the pseudo-autosomal and adjacent Y-specific region. **A** The Y chromosome with the pseudoautosomal region in *black* and Yq heterochromatin *hatched*. Intervals 1–7 are according to Vergnaud et al. (1986), with interval 4 B harboring the centromere (Annerén et al. 1987). Sizes of intervals are arbitrary and do not imply actual physical size. Reference markers are given for each interval, with markers selected for a framework map of the Y chromosome (Weissenbach and Goodfellow 1991) in *boldface*. **B**. Physical map of the pseudoautosomal and adjacent Y-specific region. The pseudoautosomal map is compiled from data in Petit et al. (1988), W.R.A. Brown (1988) Rappold and Lehrach (1988), and Rappold et al. (1991), while the male recombination frequencies of the pseudoautosomal loci with sex are from Rouyer et al. (1986), Goodfellow et al. (1986), Page et al. (1987a), Petit et al. (1988), Gough et al. (1990), and Johnson et al. (1991). The positions of SRY (Sinclair et al. 1990), RPS4Y (Fisher et al. 1990b), and ZFY (Page et al. 1987c) are indicated within interval 1, which is subdivided according to Page et al. (1987c). Note fivefold expansion of scale in the Y-specific relative to the pseudoautosomal region. As in **A**, reference markers for a framework map of the Y chromosome are in *boldface*

in contrast to the mouse (Soriano et al. 1987; Keitges et al. 1987), no double crossing over has so far been observed in this region, a finding which awaits an explanation, if it is consistent.

The occurrence of an obligatory chiasma with the consequence of a 50% recombination frequency is exceptional for such a small chromosomal segment as the pseudoautosomal region. Estimates of its length based on long-range restriction maps are 2.6–3 Mb (Petit et al. 1988; W.R.A. Brown 1988; Rappold and Lehrach 1988). In comparison, in female meiosis the recombination frequency between the most telomeric marker DXYS14 and MIC2 is only 3% (Weissenbach et al. 1987). To explain this phenomenon, Weissenbach et al. (1987) have put forward the hypothesis that minisatellite sequences distributed within the pseudoautosomal region represent hot spots of recombination. Indeed, the loci DXYS14, DXYS15, DXYS17, and DXYS20, which were used to map the pseudoautosomal region, are associated with hypervariable minisatellites. Another possibility is that in the male, a specific factor interacts with sequences which in cis promote recombination (Ellis and Goodfellow 1989).

The pseudoautosomal region must be delimited by an effective boundary to prevent recombination events in the adjacent sex-determining segment of the Y chromosome. The boundary may be expected to be a special structure. However, it turned out that sequence divergence starts at the insertion site of an Alu element. This Alu sequence is found in the Y only, but not in the X chromosome at this position (Ellis et al. 1989; Fig. 2). Homology between the sex chromosomes is complete from the telomeres to the 303 bp Alu element, followed by a stretch of 220 bp with 78% sequence identity. Proximal to that, X and Y sequences diverge definitely. Thus, the pseudoautosomal boundary region is marked by an Alu element in the Y and includes the partially homologous 220 bp segment on both X and Y chromosomes. It was designated PABX and PABY for *pseudoauto-somal boundary X* and *Y*, respectively (Weissenbach et al. 1989). It appears that recombination peters out within the boundary region, and therefore complete sequence identity was not maintained. In a population study it was shown that PABX is highly, PABY only weakly polymorphic, indicating a more recent origin of the Y boundary compared with the X boundary (Ellis et al. 1990a).

It is assumed that, originally, the pseudoautosomal boundary was simply the point of transition from homology to divergence at the proximal end of this 220-bp segment, and that the Alu element was inserted later during evolution. This view is supported by studies on primates showing that in the great apes the Alu element is also present at the boundary, while it is absent in some old-world monkeys. Thus, the Alu element must have been inserted at the preexisting boundary during hominoid evolution less than 25 million years ago (Ellis et al. 1990b).

There is evidence that the present-day pseudoautosomal region of the human Y is only part of a considerably more extended segment of homology between the sex chromosomes which was disrupted by a pericentric inversion (Yen et al. 1988), as discussed below.

3.2 X-Y Homologies Outside the Pseudoautosomal Region

If we now enter the discussion on the differential segment of the Y chromosome, we may assume that we are dealing with sequences that have no homologies to other chromosomes and are therefore Y specific. However, as is to be expected from the foregoing considerations about the evolution of the human Y chromosome, most of the differential segment may originate from other parts of the genome; therefore, homologous sequences outside the Y should exist, and this is exactly what has been found. The question then follows whether these homologies interfere with the functional requirement of an effective isolation of the differential segment.

3.2.1 Evidence for a Region of Pseudoautosomal Origin

It should be mentioned first that the precise regional location of genes and anonymous DNA sequences on the differential segment of the Y chromosome is still somewhat ambiguous. A map of reference markers has been constructed which can serve to define positions of loci (Weissenbach and Goodfellow 1991; Fig. 3). However, this reference map is still very imprecise, in particular for the differential segment. A more detailed deletion map of the Y chromosome long arm has been provided by Bardoni et al. (1991) and Yen et al. (1991) and will be of great help for further analysis.

However, not many loci have yet been assigned within the framework of these maps, and their positions are only roughly known in terms of either the long or short arm or the centromeric region of the Y. The present state of the map nevertheless allows for tentative conclusions on conserved linkage groups and structural changes showing that the Y chromosome represents a patchwork of segments and sequences of heterogeneous origin.

Concerning the genes listed in Table 1, KALIG-1Y/ADMLY has been assigned to Yq11.21, as has STSP. Their homologues on the X chromosome are both located on Xp22.3 about 7–9 Mb proximal to telomere of Xp (Petit et al. 1990), and KALIG-1X/ADMLX is 1.5 Mb proximal to STS (Franco et al. 1991; Legouis et al. 1991). Therefore, KALIG-1Y/ADMLY and STSP may be at a similar distance to each other on the Y. AMGX maps within Xp22.1–p22.31 proximal to the STS locus, and AMGY was tentatively mapped near the centromere on Yq11 (Lau et al. 1989a). According to Nakahori et al. (1991b), however, AMGY is located on proximal Yp, as indicated by the study of an XX male case. This location on Yp11.2 was confirmed by Salido et al. (1992), and the original assignment to Yq11 by Lau et al. (1989) was newly interpreted by redefining the aber-

Table 1. Genes, pseudogenes, and phenotypic characters mapped on the Y chromosome and their X homologues

Y chromosome		X chromosome		Comments
CSF2RA	Yp11.3	CSF2RA	Xp22.33	Pseudoautosomal
MIC2	Yp11.3	MIC2	Xp22.33	Pseudoautosomal
SRY	Yp11.3			= TDF
RPS4Y	Yp11.3			Transcribed gene
		RPS4X	Xq13.1	Escapes X-inactivation
ZFY	Yp11.3			Transcribed gene
		ZFX	Xp21.3–p22.1	Escapes X-inactivation
TSPY	Yp proximal			Transcribed gene
GBY	Yp11.1–qter			Phenotypic evidence
AMGY	Yp11.2			Transcribed gene
		AMGX	Xp22.1–p22.31	Escapes inactivation (?)
ADMLY	Yq11.21			Pseudogene (?)
		ADMLX	Xp22.3	Escapes X-inactivation
STSP	Yq11.21			Pseudogene
		STS	Xp22.32	Escapes X-inactivation
ASSP6	Ycen–q11			Pseudogene
ACTGP2	Yq11			Pseudogene
GCY	Yq11			Phenotypic evidence
H-Y (H–Yc)	Yq11.2			Phenotypic evidence
AZF	Yq11.23			Phenotypic evidence
RVNP2	Y			

rant Y chromosome studied by these authors; this interpretation is compatible with the location of AMGY on Yp (see Salido et al. 1992). Assuming STS, KALIG-1/ADML, and AMG to have been part of a former considerably larger pseudoautosomal region which was subsequently disrupted on the Y chromosome by a pericentric inversion, a secondary and smaller pericentric inversion has to be postulated in order to explain the Yp location of AMGY (Fig. 2).

There is an indication that the ancestral pseudoautosomal region also includes some further loci which are present on proximal Yp in addition to AMGY (Petit et al. 1990). A DNA probe recognizing a family of interspersed repetitive sequences, DXYZ2 (so-called STIR elements, see below under Sect. 4.3), located on Xp22.3 and on various sites of the Y chromosome, was subcloned for the purpose of isolating other members of this family. Among the loci detected in this way are two which show cross-homology on Xp22.3 and Yp11.2, locus DXS431 corresponding to DYS137 and locus DXS432 corresponding to DYS138. On the X chromosome, these loci are located between PABX and STS, and they may therefore belong to the ancestral pseudoautosomal segment. If so, they should have been transposed from proximal Yq to proximal Yp by a secondary rearrangement. Finally, ASSP6 was tentatively assigned to Ycen–q11 (T.S. Su et al. 1984), and there is a closely related gene ASSP4 on Xp22–pter, but also another one, ASSP5 on Xq22–q26 (T.S. Su et al. 1984).

The dissection of the Yq11 region by Bardoni et al. (1991) and Yen et al. (1991) is an impressive confirmation of the existence of a conserved linkage group on both Xp22.3 and Yq11.2. These authors defined the order of several loci, including STSp which map within segments Yq11.21–q11.22, and these loci have homologous sequences on Xp22.3 in the same relative order, comprising about 5 Mb. On the X chromosome, this linkage group is immediately adjacent to the pseudoautosomal region, confirming the assumption that it was included in a formerly much larger pseudoautosomal region which was subsequently disrupted on the Y by a pericentric inversion. If so, the gene order on the Y should be inverse to that on the X, and this indeed seems to be the case (Bardoni et al. 1991; Yen et al. 1991). Similarly, the order of the above-mentioned genes or pseudogenes on the Y appears to be AMGY – Ycen – KALIG-1Y/ADMLY – STSP – Yqter, and on Xp to be Xpter – STS – KALIG-1X/ADMLX – AMGX – Xcen (Fig. 2 B). The position of the Y centromere is due to a second pericentric inversion, as mentioned above.

Studies on STS indicate that this gene may still behave pseudoautosomally in prosimians (Yen et al. 1988; Schempp and Toder 1992) and definitely does so in the mouse (Keitges et al. 1985). Therefore, an inversion must have occurred in the human ascendence when the higher primates diverged from the prosimians about 40 million years ago. This assumption fits well with estimates based on sequence similarity between STS and STSP, which arrived at 40 million years (Yen et al. 1988). For the KALIG-1 gene, as for STS, no dosage difference between male and female DNA was detected on Southern blots in prosimians (lemurs) thus indicating pseudoautosomal behavior, while in higher primates its dosage depends on the number of X chromosomes (Franco et al. 1991). This is a further indication for the occurrence of an inversion disrupting an originally coherent pseudoautosomal segment.

Since the proximal long arm of the Y shares homology with the distal X short arm proximal to PABX, it has been proposed that pairing and crossing over may occasionally occur in male meiosis, resulting in X-Y translocations (Ballabio et al. 1989). There are a number of reports giving evidence for this phenomenon. In four cases analyzed by Geller et al. (1986) the breakpoints were in Xp22.3 and Yq11 and were assumed to be close to STS and STSP, respectively (Yen et al. 1988). Further examples are presented by Ballabio et al. (1989) and Yen et al. (1991), including one translocation in which sequencing of the breakpoints demonstrates the occurrence of homologous recombination. Bardoni et al. (1991) and Franco et al. (1991) localized the breakpoints of three Xp22.3/Yq11.2 translocations within a homologous segment, including the KALIG-1Y/X loci. These findings favor the view that these translocations originate from meiotic pairing and crossing over.

Further evidence for the existence of a region of pseudoautosomal origin comes from functional aspects and sequence information. The KALIG-1 gene on the X chromosome escapes inactivation (Franco et al. 1991), and this is partially so with the STS gene (Migeon et al. 1982). To avoid a gene dosage difference between the sexes, the homologous genes on the Y should also be active, as is the case with the genes in the pseudoautosomal region on distal Yp/Xp, MIC2 and CSF2RA. For the KALIG-1Y gene evidence as to its functions is still lacking; however, it is assumed to be inert because it cannot complement a defect in KALIG-1X, and this is clearly the case with the STSP gene. However, the sequences of these genes are quite similar to those of their X homologues, and they may still have been functional in their more recent past (Weissenbach and Goodfellow 1991). In the case of the STS gene, we may have an example of the transitional stage from activity to inactivation, since its activitiy is reduced on the inactivated X chromosome (Migeon et al. 1982) while its counterpart on the Y has already become a pseudogene. For AMGX, the problem of X-inactivation is not settled yet. As discussed unter Sect. 5.7, this locus may escape X-inactivation, but its expression, if on the inactivated X, is markedly reduced (Salido et al. 1992), as in the case of STS. The homologue on the Y chromosome, AMGY, is transcribed, but the amount of gene product was estimated to be only 1/10 of that of AMGX (Salido et al. 1992). Under the functional aspect it can therefore be concluded that all three genes, KALIG-1, STS, and AMG, behaved pseudoautosomally up to the recent past.

In summary, there is compelling evidence for the disruption of an originally contiguous, large pseudoautosomal region by a pericentric inversion on the Y chromosome having occurred during early primate evolution. The order of loci within this region on Yq is retained, but it is the inverse of the order on the X chromosome (Fig. 2 B). The homology is such that meiotic pairing and recombination can still occur, resulting in rare X-Y translocations. Finally, the degree of sequence identity to their X homologues of the few genes assigned or tentatively assigned to this region is compatible with a pseudoautosomal origin of these genes. By secondary rearrangements, part of this ancestral pseudoautosomal segment appears to have been transposed to the proximal short arm of the Y chromosome.

3.2.2 Other Regions of X-Y Homology

Various proposals have been made to categorize the loci showing X-Y homology and to assign these loci to different regions of the Y chromosome (Koenig et al. 1985; Affara et al. 1986b; Bickmore and Cooke 1987; Bardoni et al. 1991). Following and modifying the classification of Bickmore and Cooke (1987), we suggest dividing the different X-Y homologous sequences into five main categories according to their degree of sequence identity and their X or Y status in other primates. In this scheme, category I encompasses the sequences from the (still functional) pseudoautosomal region on distal Xp and Yp just discussed which are X-Y identical (category III in Bickmore and Cooke 1987). Category II contains sequences from distal Xq and Yq, which are also X-Y identical but are X-specific in higher primates (grouped into category I by Bickmore and Cooke 1987). Category III is defined by sequences from Xq and Yp with over 98% identity, which are again X only in higher primates (category I of Bickmore and Cooke 1987). Category IV (category II of Bickmore and Cooke 1987) encompasses the sequences from the presumptive ancestral pseudoautosomal region on Xp and proximal Yq discussed above, showing a sequence identity in the range of 80%–95%, which are X-Y homologous in great apes and old-world monkeys but are X only in new-world mon-

Fig. 4 (page 169). XY homologous regions on human sex chromosomes. Five categories of X-Y homologous segments are indicated, classified according to the degree of sequence conservation. Their positions on the G-banded X and Y chromosomes are given

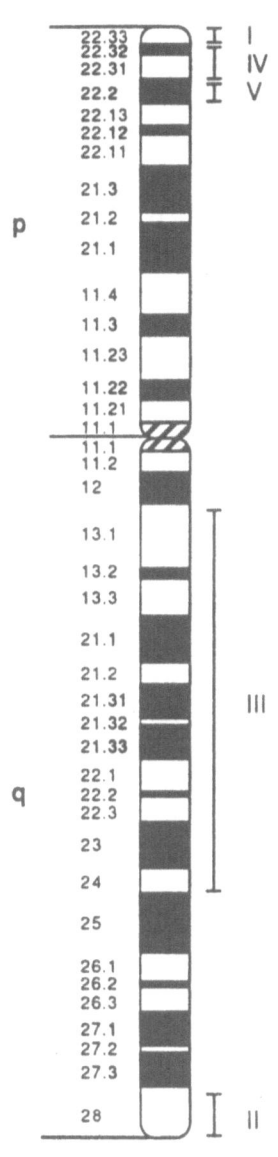

X

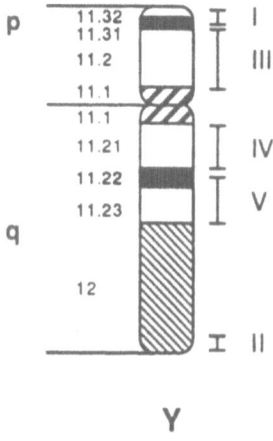

Y

Category	Location	Characteristics (type loci or probes)	Sequence identity
I	Xp22.33 ; Yp11.32	Pseudoautosomal region (DXYS14, MIC2)	100 %
II	Xq28 ; Yq12	X-only in great apes (DXYS61, DXYS64)	100 %
III	Xq13-Xq24 ; Yp11	X-only in great apes (DXYS1, DXYS5)	98-99 %
IV	Xp22.3 ; Yq11.21	X-Y homology in great apes and old world monkeys; X-only in new world monkeys (STS/STSP, AMGX/AMGY)	80-95 %
V	Xp22.2 ; Yq11.22 -Yq11.23	weak X-Y homology (71-7A, Fr25II)	?

keys. Finally, category V contains sequences from Xp and distal Yq with a lower degree of sequence conservation. For orientation, these five X-Y homologous regions are displayed schematically in Fig. 4.

Corresponding to the pseudoautosomal region (category I) on the telomeric ends of the short arms of X and Y, a small segment of complete sequence identity (category II) has been identified in terminal position on the long arms. This segment is at present defined by only two loci, DXYS61 (probe Y2:13) and DXYS64 (probe St35-239), which are less than 130 kb apart on Xq28 (Arveiler et al. 1989) and map to distal Yq12 (Bardoni et al. 1991; Pedicini et al. 1991). DXYS61 shows X-Y homology extending for at least 50 kb and no difference in 1200 bp of sequence from corresponding X and Y clones (Cooke et al. 1984; Bickmore and Cooke 1987). Interestingly, DXYS64 was still present in an individual showing reduced size of the Yq12 heterochromatin, a region of considerable interindividual size variation (Pedicini et al. 1991). If confirmed in more cases, this variation must be due to interstitial changes. This terminal homologous segment of man is found only on the X chromosome in higher primates; therefore, it must have been transposed onto the Y during recent evolution. According to Chandley et al. (1984), quoted in Pedicini et al. (1991), the terminal parts of the long arms of X and Y are frequently associated during meiosis in early pachytene. Thus, recombination in this region may occur, and this may be the reason for the conservation of complete homology. If so, we are dealing with still another pseudoautosomal region.

The first single-copy sequence found with homologues on both sex chromosomes is pDP34 (pDP31), defining locus DXYS1 (Page et al. 1982, 1984). Its X homologue DXYS1X has been assigned to Xq21.3, and its Y homologue DXYS1Y to proximal Yp11. In the nomenclature of Vergnaud et al. (1986), who constructed a deletion map of the Y chromosome dividing it into seven deletion intervals (see Fig. 3 A and p. 190), DXYS1Y is located in interval 4A. Subsequently, many other anonymous sequences have been identified, all located within Xq13 q21 and Yp11, mostly in intervals 1C and 2 of Vergnaud et al. (1986). Together they define category III of the X-Y homologous sequences and include the loci DXYS5 (Guellaen et al. 1984), DXYS2-4 and DXYS6-9 (Geldwerth et al. 1985), DXYS12 (Koenig et al. 1985), DXYS21 and DXYS30-32 (Affara et al. 1986b), and DXYS49 (Waibel et al. 1987; a complete listing is found in Weissenbach and Goodfellow 1991). The region on Yp from which these loci derive has been estimated to comprise about one quarter of the Y euchromatin (Koenig et al. 1985), or about 10 Mb. The sequence identity between the corresponding X and Y loci has been estimated from comparative restriction mapping to be about 99% for DXYS1 (Page et al. 1984) and about 98% for DXYS12 (Koenig et al. 1985), and from analysis of

1350 bp of sequences from corresponding X and Y clones to be 98.5% for DXYS49 (J. Zehender and G. Scherer, unpublished data). Since in all cases analyzed the homologous sequences are restricted to the X chromosome in the great apes, it is assumed that they were transposed from Xq to Yp during recent evolution, after the separation of the great apes from the human branch. Because, as mentioned above, these X-Y homologous sequences are found in intervals 1C and 2 (e.g., DXYS5, DXYS49), and in interval 4A (e.g., DXYS1) on Yp, separated by strictly Y-specific sequences defining interval 3 (Vergnaud et al. 1986; Fig. 3 A), two such transposition events may have occurred at about the same time in human evolution. Alternatively, they may have been moved en bloc onto Yp, to become separated later on by a paracentric inversion. Interestingly, in a few XX males (see Sect. 7.1) Y sequences from intervals 1C, 2, and 4 are present in the absence of some interval-3 sequences (Affara et al. 1987; Brøndum Nielsen et al. 1988; Scherer et al. 1989b), pointing to the existence of such a paracentric inversion in extant Y chromosomes. Corresponding results have been obtained for some Yp-deleted females with XY gonadal dysgenesis (Disteche et al. 1986b; G. Scherer, unpublished data).

The euchromatic part of Yq contains two categories of X-Y homologous sequences: loci like STSP belonging to the presumptive ancestral pseudoautosomal region comprising category IV on proximal Yq, band Yq11.21 (region I in Bardoni et al. 1991), and various loci of weak X-Y homology defining category V located more distally in bands Yq11.22 and Yq11.23 (region II in Bardoni et al. 1991). Loci belonging to category V were identified first by Kunkel et al. (1983) with probe 71-7A defining locus DXS69 on Xp22.2 and various loci on Yq11.22–q11.23 (Bardoni et al. 1991). Similarly, probe Fr25II (Scherer et al. 1989b; Bardoni et al. 1991) detects a locus on Xp22.2 and various loci again on Yq11.22–q11.23. The locus DXS29 may also belong to this category (Affara et al. 1986b), but showing no duplications. The X-Y homology is detected only under low-stringency hybridization conditions, pointing to a lesser degree of sequence conservation. It is to be assumed that transposition of these sequences from the X to the Y occurred earlier in evolution than the transposition events discussed above, and that these loci were duplicated afterwards by rearrangements within the Y chromosome.

A last category of X-Y homologous loci is represented by the pseudoautosomal locus DXYS77 which has a duplicate, DYS148, of 95% sequence identity on proximal Yq (Fisher et al. 1990a). It has been shown that other sequences in the vicinity of DXYS77 and DYS148, comprising several kilobases in length, are of similar sequence identity, and a larger block of DNA was therefore included in this interarm duplication. From studies of these sequences in primates, it can be derived that the duplica-

tion event took place before the divergence of the human species and the
great apes (Fisher et al. 1990a).

Finally, it must be mentioned that many anonymous probes with X-Y
homology also detect autosomal loci, adding still another level of complexity to the evolutionary history of Y chromosomal sequences (e.g., Affara
et al. 1986b; U. Müller et al. 1986a, b; Vergnaud et al. 1986; Bardoni et al.
1991).

4 Repetitive Sequences

Though not restricted exclusively to male sex determination/differentation,
the human Y chromosome carries very few functional genes for its size. It
has been estimated to consist up to 70% of highly repetitive DNA (Cooke
et al. 1983) which is not transcribed; a more realistic figure seems to be
30%–40%. There are numerous other reiterated sequences, either in small
clusters or intercalated at various locations as families of homologous
sequences. While some repeat clusters were found to be present also on the
Y chromosomes of various primate species (see, e.g., U. Müller 1987),
others do not have homologues on the Y chromosomes of primates closely
related to the human species, but rather on the X chromosome or on
autosomes; therefore, their Y location must be of recent origin (reviewed
by Smith et al. 1987). It is assumed that the sequences were transposed to
the human Y chromosome at various periods during its evolution after di-
verging from the great apes, and that some of them became selectively amplified, resulting in tandem repeats. Due to the meiotic isolation of the
nonpairing segment of the Y, variation of repetitive sequences was largely
unlimited, and this variation is still reflected in present-day human Y chromosomes. This variation, including frequent inversions, points to a "linear
instability" of the nonpairing region of the Y chromosome (Affara et al.
1986a).

Principally, repetitive DNA can be classified in tandem clustered repeats, short interspersed repetitive elements (SINEs), and long interspersed
repetitive elements (LINEs), and their occurrence on the Y chromosome
was reviewed by Smith et al. (1987). Among SINEs, Alu repeats are present in considerable number on the Y but seem to have significantly diverged from the average Alu repeats found on other chromosomes. Only a
few homologies are detected unter high-stringency conditions using average Alu probes (Wolfe et al. 1984), but under relaxed conditions their
number increases considerably (Burk et al. 1985). LINEs are also abundant
on the Y chromosome, and they are indistinguishable from those of other
chromosomes, showing no chromosome-specific organization. Their fre-

quency corresponds to the size of the Y chromosome as compared with other chromosomes (Schmeckpeper et al. 1981). There are various clusters of tandem repeats, the most prominent of which will now be discussed separately.

4.1 C-Heterochromatin

The human Y chromosome exhibits a considerable interindividual size variation (Cohen et al. 1966). After the discovery by Zech (1969) of a brightly fluorescing segment on distal Yq using quinacrine derivatives for chromosome staining, it was shown that the size variation is due exclusively to varying amounts of the fluorescing part (Bobrow et al. 1971), which may even be entirely lacking without any phenotypic effect (Borgaonkar and Hollander 1971). Cytogenetically, this fluorescing segment is part of constitutive heterochromatin as defined by the C-banding method (Arrighi and Hsu 1971). The C-heterochromatin extends proximally beyond the fluorescing segment, and if it is totally lacking, azoospermia results (Tiepolo and Zuffardi 1976).

Restriction enzyme analysis revealed this heterochromatic segment to consist of tandem repeats of DNA, showing a male-specific pattern in Southern blots. After *Hae*III digestion, two fragments of 3.4 kb and 2.1 kb are seen in male but not in female DNA (Cooke 1976; Bostock et al. 1978). The 3.4-kb repeat (designated DYZ1) constitutes about 20% (4000 copies) and the 2.1-kb repeat (designated DYZ2) about 10% (2000 copies) (Cooke 1976) of the Y chromosome, assuming a size of 60 Mb (Morton 1991). The complete nucleotide sequence of a 3.56-kb *Eco*RI fragment comprising one DYZ1 unit revealed an uninterrupted tandem array of the pentanucleotide TTCCA and variants thereof (Nakahori et al. 1986). The 2.1-kb DYZ2 *Hae*III repeat is included in a larger 2.47-kb repeat that consists of a highly AT-rich and a GC-rich region with a nearly complete Alu element (Frommer et al. 1984). Size variation of the heterochromatic part correlates with the amount of the 3.4-kb DYZ1 repeat (McKay et al. 1978), and of the 2.1-kb DYZ2 repeat as well (Schmid et al. 1990), thus clearly locating these repeats into the heterochromatic segment. By in situ hybridization to metaphase chromosomes it has been shown that both repeats, DYZ1 and DYZ2, are spread throughout the heterochromatic region (Bostock et al. 1978; Y.F. Lau 1985; Schmid et al. 1990). Furthermore, in situ hybridization studies revealed that the DYZ1 repeat has some sequence homology to the heterochromatic segments of chromosomes 9 and 15 (Bostock et al. 1978; Schwarzacher-Robinson et al. 1988). Under hybridization conditions requiring 80%–85% sequence identity, hybridization was still significant to chromosome 9 heterochromatin, but scarce to that of chromosome 15 (Schwarzacher-Robinson et al. 1988). Sequences related to the 2.1-kb repeat are also present on other chromosomes concentrated at the telomeres (Cooke et al. 1982).

The 2.1-kb repeat has been studied in higher primates, including the gorilla, chimpanzee, and gibbon (Cooke et al. 1982). While hybridization was not detectable in the gibbon, related sequences are present in the other two species. However, no sex-specific pattern was found. It is hypothesized that the 2.1-kb fragment was amplified on the Y chromosome during human evolution, and that it was maintained en bloc due to its isolation.

Sequences related to the 3.4-kb repeat are also contained in the genomes of the gorilla and the chimpanzee, but again no sex difference was detected (Cooke et al. 1982). Clusters of sequences related to the 3.4-kb repeat are widely distributed in the human genome (Burk et al. 1985); this should also be the case in the close human relatives, and its amplification on the Y is therefore assumed to be an event which happened to take place only during human evolution.

4.2 Centromeric Alphoid Repeats

The centromeric alphoid satellite DNA locus of the Y chromosome, designated DYZ3, consists of tandem repeated monomeric segments of 171 bp (Wolfe et al. 1985), the fundamental repeat unit size of all alpha satellites, organized in 5.7- or 6.0-kb units which are themselves tandem repeated in a single large, uninterrupted array of variable length (Tyler-Smith and Brown 1987). The length polymorphism varies between 250 kb and 1200 kb. Most males studied can be classified into two size-groups of the DYZ3 repeat, either in the range of 250–450 kb or in the range of 800–1200 kb (Wevrick and Willard 1989; Oakey and Tyler-Smith 1990). This bimodal distribution, which seems to be unique to the Y chromosome, was interpreted as having resulted from a mitotic recombination event creating two unequally sized arrays early in the evolution of this DNA family, which by necessity cannot undergo meiotic recombination between homologues (Wevrick and Willard 1989). The DYZ3 repeat size has been used, in combination with other hypervariable loci, for Y chromosome haplotyping, suggesting descendence of most Eurasian men from one of two males (Oakey and Tyler-Smith 1990).

4.3 STIR Elements

Subtelomeric interspersed repeated (STIR) elements are a family of sequences originally detected in the pseudoautosomal region and designated DXYZ2 (Simmler et al. 1985). On the Y chromosome they were found to be dispersed over the entire pseudoautosomal region (W.R.A. Brown 1988; Petit et al. 1988), but they are also present in the proximal regions of the long and short arms believed to be a segment of the ancestral pseudoautosomal region (see above) (Petit et al. 1990).

STIRs are also specifically located in subtelomeric regions of autosomes. Therefore, the nomenclature was changed and this family of repeats is now called DNF28 (Petit et al. 1990). The sequences of a number of STIRs have been determined, including four loci from the pseudoautosomal region and nine elements from four autosomal loci (Rouyer et al. 1990). A monomeric element comprises about 350 bp, and in the pseudoautosomal region they are organized in tandem multimeres of two or four elements. Based on sequence differences, two types were distinguished, and the multimers include both types. In contrast, on autosomes the STIR elements exist as monomers of one type only. Interestingly, on the X chromosome short am STIRs are found also proximal to the present pseudoautosomal region (Petit et al. 1990), confirming the assumption that a larger pseudoautosomal region existed in the past, as discussed above. This finding also corresponds to the observation of STIR elements in the paracentric regions of the Y as part of the ancestral pseudoautosomal region.

Linkage data revealed greatly increased recombination frequencies, especially in male meiosis, in the subtelomeric regions of all chromosomes; therefore, it is speculated that the STIR elements may serve the function of promoting the initiation of pairing at meiosis (Rouyer et al. 1990).

5 Genes and Pseudogenes

Up to the present, 12 genes or pseudogenes of the Y chromosome have been identified and cloned, seven of which were shown to be expressed genes, while the others are either known or suspected to be pseudogenes. These loci are listed in Table 1 (see p. 165) together with homologous loci on the X chromosome. In addition, genes postulated to exist on the basis of indirect evidence (see Sect. 6) are also included in the Table. These loci will now be discussed.

5.1 Colony-stimulating Factor 2 Receptor Alpha (CSF2RA)

A cDNA of the receptor for the granulocyte-macrophage colony-stimulating factor has been cloned (Gearing et al. 1989), and by in situ hybridization it was shown to map to the terminal shorts arms of the X and Y chromosomes (Gough et al. 1990). Pseudoautosomal inheritance was demonstrated by segregation analysis of *Hind*III polymorphic fragments of this gene. It shows a male recombination frequency of about 20% with respect to the sexual phenotype (Gough et al. 1990) and maps about 1.2–1.3 Mb from Ypter (Rappold et al. 1991), between DXYS15 and DXYS17 (Fig. 3 B). Both X- and Y-located homologues of CSF2RA were suggested to be expressed because they may be involved as recessive oncogenes in the generation of acute myeloid leukemias of the M2 subtype. In this condition, 25% of cases have lost either an X or the Y chromosome, and a significant proportion of patients are unresponsive to the granulocyte-macrophage colony-stimulating factor (quoted in Gough et al. 1990).

5.2 MIC2

The MIC2 gene was identified by the monoclonal antibody 12E7, which was raised against human leukemic T cells (Levy et al. 1979). The symbol MIC2 refers to *m*onoclonal antibody no. 2 of the *I*mperial *C*ancer Research Fund (Shows and McAlpine 1982). The gene is expressed on the surface of all somatic cells tested (Goodfellow 1983) and was shown by somatic cell hybrid studies to be located on the tip of the X chromosome short arm and to have a homologue on the Y chromosome (Goodfellow et al. 1983). Upon cloning of a cDNA for MIC, the MIC2X and MIC2Y loci were shown to be closely related, if not identical (Darling et al. 1986), and by DNA in situ hybridization both loci were shown to be located in the pseudoautosomal region of the X and Y chromosomes (Buckle et al. 1985). Consequently, MIC2X escapes X-inactivation (Goodfellow et al. 1983, 1987b). The recombination fraction of MIC2 with respect to sex was determined to be between 2.5% (Goodfellow et al. 1986) and 0% (Johnson et al. 1991). Correspondingly, MIC2 maps very close to the pseudoautosomal boundary (Petit et al. 1988; Fig. 3 B). The function of the MIC2 antigen is not known.

5.3 SRY

The symbol SRY stands for *sex*-determining *region Y* gene (Sinclair et al. 1990). The SRY gene was identified in a systematic search for the gene encoding the testis-determining factor (TDF). There is strong evidence that SRY is indeed identical with TDF, and we will therefore discuss this gene in the section below on sex determination.

5.4 Ribosomal Protein S4 (RPS4)

Ribosomal protein S4 is one of approximately 80 different protein components of the ribosome, and it is found among the some 33 proteins of the smaller ribosomal subunit. The search for expressed DNA sequences within the testis-determining region of the Y chromosome succeeded in identifying a transcript encoding a 263-amino acid protein, which by sequence comparison turned out to be largely homologous to the already-known ribosomal protein S4 of the rat (Fisher et al. 1990b). Though this gene is located proximally, outside the pseudoautosomal region (Fig. 3 B), an X-linked homologue was found, and it was mapped by in situ hybridization and analysis of somatic cell hybrids to the long arm of the X chromosome, within the region Xq13.1 near the inactivation center (XIC). Sequence identity in the coding region between RPS4Y and RPS4X is 82%, at the protein level 93%. Surprisingly, RPS4X escapes X-inactivation, and it is so far the only gene on the X long arm showing this property, except for a sequence identified in XIC named XIST, which, however, is inactivated on the active X chromosome (C.J. Brown et al. 1991a, b). Escape from inactivation and the occurrence of a functional homologue on the Y chromosome make RPS4 a candidate for a so-called anti-Turner gene, because in the XO condition (Turner's syndrome), it is present in single dose only and this may cause haploinsufficiency. However, in patients with an isochromosome for the X chromosome long arm, at least a triple dose of the gene is expressed, and they nevertheless exhibit Turner's syndrome (Just et al. 1992).

5.5 ZFY

A strategy to identify the gene for TDF was to narrow down its location by analyzing XX males carrying the testis-determining region of the Y at the tip of one of their X chromosomes and XY females having a deletion in this region (see Sect. 7). The region of overlap of the deletion and the translocated segment is expected to harbor the gene responsible for male development. By this approach, Page et al. (1987c) defined a region of 140 kb, and within this region they detected a sequence with an open reading frame corresponding to 404 amino acids. The deduced polypeptide is a zinc finger protein with consecutive repeats of 13 Cys-Cys/His-His fingers similar to a variety of transctiption factors like TFIIIA from *Xenopus* oocytes. Therefore, the gene was named ZFY, for *zinc finger protein Y*-linked (Page 1988). The DNA probe detecting ZFY also cross-hybridizes with a segment on the X chromosome (Page et al. 1987c), named ZFX, and by in situ hybridization ZFX was mapped to the interval Xp21.3–p22.1 (Müller and Schempp 1989; Affara et al. 1989; Page et al. 1990a). The complete ZFY protein encompasses 801 amino acids (Lau and Chan 1989; Palmer et al. 1990), while for ZFX, protein isoforms of 804 and 575 amino acids resulting from alternatively spliced transcripts were described (Schneider-Gädicke et al. 1989a). The ZFY and ZFX proteins are closely related, each composed of an acidic domain, a small basic domain, and 13 Cys-Cys/His-His zinc fingers – a combination of features reminiscent of transcription activators. This relationship is reflected in a high sequence conservation, which is highest in the zinc finger domain, showing 98% sequence identity (Schneider-Gädicke et al. 1989b).

Both ZFY and ZFX are expressed in the adult human testis, but also in many other tissues (Schneider-Gädicke et al. 1989b; Lau and Chan 1989). Studies on transcriptional activity of ZFX using somatic cell hybrids containing the inactivated human X chromosome on rodent background revealed this gene to escape X inactivation (Schneider-Gädicke et al. 1989b). Both ZFX and ZFY were found to be present on the sex chromosomes of all eutherian mammals examined (Page et al. 1987c). In the mouse, the situation is more complex with a total of four genes: two on the Y, Zfy-1 and Zfy-2; one on the X, Zfx; and another, Zfa, on chromosome 10 (Mardon et al. 1989; Nagamine et al. 1989). In addition, in the mouse Zfx undergoes inactivation (Ashworth et al. 1991) In contrast to eutherians, in marsupials the sequences homologous to ZFX/ZFY are of autosomal location (Sinclair et al. 1988), as is the case in nonmammalian vertebrates like the chicken (Page et al. 1987c). ZFX shows stronger conservation in mammals and other vertebrates than ZFY, and it is assumed that it was originally an autosomal gene which became transposed to the sex chromosomes during eutherian evolution.

Studies on the spatiotemporal expression of ZFY/ZFX and their mouse homologues, and speculations on their possible function, have recently been reviewed by Koopman et al. (1991a). Based on findings in the mouse, a role in spermatogenesis was ascribed to Zfy-1 and Zfy-2 (Mardon et al. 1989; Nagamine et al. 1989; Koopman et al. 1989). However, since both

ZFX and ZFY are ubiquitously expressed in human somatic cells, these genes may serve a more general function in the organism. If these zinc finger proteins are transcription factors, the next question to be answered will be which genes they control.

5.6 TSPY

This symbol refers to "*t*estis *s*pecific *p*rotein *Y*-encoded" (Weissenbach et al. 1989). The gene in question was identified in an attempt to detect transcribed DNA sequences specific for the Y chromosome (Arnemann et al. 1987). Only a single transcript was detected, and the gene was characterized further (Arnemann et al. 1991). It appears to be intronless, and the amino acid sequence was deduced to have 258 residues. TSPY was mapped to the proximal part of the Y chromosome short arm. No homologous sequence was found on other chromosomes, but an apparent pseudogene to TSPY was identified within the same region of the Y chromosome, showing DNA sequence identity of about 92%. TSPY is expressed in early spermatids, but in no other tissues tested. It is speculated, therefore, that it may serve a function in morphological sperm differentiation (Arnemann et al. 1991). Male specificity of a segment homologous to TSPY was also detected in the chimpanzee and the cow, but it has not yet been found in some other mammalian species examined, and the sequence possibly diverged during evolution. Nevertheless, its conservation points to some basic function.

5.7 Amelogenin Y Gene (AMGY)

Amelogenin is the predominant protein of tooth enamel. Using a cDNA probe for amelogenin of the mouse, homologous sequences were detected in man. From studies of somatic cell hybrids, these sequences were assigned to the X chromosome in the region Xp22.1–p22.31, and to the pericentromeric region of the Y chromosome, possibly in proximal Yq11 (Lau et al. 1989a). Deletion mapping performed by Nakahori et al. (1991b), however, placed the Y-linked sequence to proximal Yp. This location has been confirmed by Salido et al. (1992), who mapped the Y-homologue to Yp11.2 by in situ hybridization. The amelogenin genes were designated AMG on the X and AMGL (AMG-like) on the Y (Weissenbach et al. 1989), and more recently AMGX and AMGY (Salido et al. 1992). For AMGX the question of whether or not this locus escapes X-inactivation can be debated. As discussed in Salido et al. (1992), the mosaic pattern of normal and abnormal enamel formation in female carriers of X-linked amelogenesis imperfecta suggests inactivation, whereas the thicker enamel found in individuals with X polysomies as compared with controls speaks in favor of a dosage effect. Possibly, AMGX undergoes inactivation only partially, similar to STS. AMGY has been shown to be transcriptionally active, though at the reduced rate of some 10% of its X homologue. This difference in activity might be due to differences in the promotor region, which shows only 80% sequence identity (Salido et al. 1992). In contrast, sequence identity of the protein-coding region of AMGX and AMGY is between 93% and 100% (Nakahori et al. 1991a; Salido et al. 1992). Homologous genes were shown to be on the X and Y chromosomes of old-world monkeys, of one new-world monkey species, and of the cow, while in other monkeys and in rodents only a single gene is present, located on the X chromosome (Nakahori and Nakagome 1989). Whether or not the Y-linked gene is expressed in these species is not known.

It is hypothesized (Nakahori et al. 1991a) that AMGY might be identical with a growth control gene, GCY, postulated to exist on phenotypic grounds (Alvesalo and de la Chapelle 1981; see Sect. 6.3). Similarly, Salido et al. (1992) suggest that TSY, a gene controlling tooth size and synonymous with GCY, may be identical to AMGY.

5.8 Kallmann's Syndrome Homologous Gene (KALIG-1Y; ADMLY)

A gene has been identified recently representing the locus for the X-linked form of Kallmann's syndrome, a condition associated with hypogonadotropic hypogonadism and anosmia caused by a defect in the embryonic migration of olfactory neurons. This gene has been designated KALIG-1 for "*Kal*lmann's syndrome *interval gene 1*" by Franco et al. (1991), and ADMLX for "*ad*hesion *m*olecule-*l*ike from the *X* chromosome" by Legouis et al. (1991). This latter designation refers to the structural similarity of the encoded 679 amino acid protein to cell adhesion molecules. ADMLX maps to Xp22.3 (Franco et al. 1991; Legouis et al. 1991), while a homologue identified on the Y chromosome, designated ADMLY, maps to Yq11.21 (Franco et al. 1991). The X-linked gene has been shown to escape X-inactivation in expression studies on somatic cell hybrids containing the inactive X (Franco et al. 1991). In contrast, the Y-linked Kallmann sequence, which from partial sequence analysis shows high similarity (95%) to the X-derived sequence, appears not to be expressed and is considered to have become a nonfunctional pseudogene during recent evolution. Y-linked sequences were detected in the chimpanzee and the macaque. Dosage analysis using DNA from male and female individuals suggests X-linkage in simians but, interestingly, a pseudoautosomal (or autosomal) location in prosimians (lemurs; Franco et al. 1991). As mentioned earlier in this article, the pseudoautosomal origin of the X- and Y-linked Kallmann genes is still reflected in recombination events between these loci in meiosis resulting in Xp22.3/Yq11.2 translocations.

5.9 Steroid Sulfatase Pseudogene (STSP)

The steroid sulfatase gene (STS) has been mapped to Xp22.32, closely proximal to the pseudoautosomal boundary (PABX), by various methods including somatic cell hybrids and the study of male patients with deletions of the distal short arm of the X chromosome exhibiting ichthyosis due to steroid sulfatase deficiency (for review, see Shapiro 1985). This gene escapes X-inactivation (Shapiro et al. 1979), but shows a somewhat reduced activity if on an inactivated X chromosome (Migeon et al. 1982). The gene has been cloned (Yen et al. 1987, 1988); its genomic extension is 146 kb, including 10 exons which encode a glycoprotein of 561 amino acids. A cross-hybridizing sequence was found on the Y chromosome (Yen et al. 1987; Fraser et al. 1987) and was mapped to Yq11.2 (Bardoni et al. 1991). Analysis of genomic clones showed that this Y homologue has large regions of extensive sequence similarity to STS but misses exons 1 and 7–10. Sequence analysis of the remaining "exons" 2–6 and of some intron-like areas revealed base substitutions, deletions, and insertions, giving a sequence similarity to STS ranging from 85% to 94% (Yen et al. 1988). Thus, the Y locus represents a true pseudogene, which was designated STSP (Weissenbach et al. 1989). Because there is no appreciable difference between the two genes in similarity of exons and introns, the entire STSP locus may have become nonfunctional at the point in evolution when it began to diverge from the STS gene, approximately 40 million years ago (Yen et al. 1988). Comparative studies indicate that the great apes and old-world monkeys also have the functional X-linked gene and, with the exception of the gorilla, the Y-linked pseudogene, while in new-world monkeys, as in the gorilla, a Y-linked pseudogene was not identified and may either have been lost or have diverged extremely from the original sequence. In prosimians, where no dosage differences between XX and XY DNA samples were observed, STS is assumed to be pseudoautosomal, as in the mouse (Yen et al. 1988; Schempp and Toder 1992).

Similar to the case of the Kallmann genes, the regions including STS and STSP sequences appear to participate in occasional meiotic pairing and crossing over, resulting in X-Y translocations (Yen et al. 1988, 1991).

5.10 Argininosuccinate Synthetase Pseudogene (ASSP6)

The functional gene for argininosuccinate synthetase (ASS) is located within the chromosome region 9q34–qter (Carrit and Povey 1979). Using a cDNA probe from this gene, multiple sequences have been detected dispersed over many human chromosomes including the X (Beaudet et al. 1982) and the Y chromosome (Daiger et al. 1982). These sequences represent processed pseudogenes without introns, and based on sequencing data of some representatives of this multigene family, they are highly homologous to the functional ASS gene (Freytag et al. 1984; Nomiyama et al. 1986). A total of 14 pseudogenes have been assigned to various chromosomes or chromosome regions, of which two are located on the X chromosome, ASSP4 on Xpter–p22 and ASSP5 on Xq22–q26, and one on the Y chromosome, ASSP6 on Ycen–q11 (T.S. Su et al. 1984). Because of the high homology of these pseudogenes to the ASS cDNA, they may have originated by a retroposon-like mechanism from ASS and become dispersed more recently. Therefore, their location in the genome should by independent of linkage groups. From the sequence difference to the ASS cDNA, a Y-located pseudogene designated ψAS-Y, probably identical with ASSP6, has been estimated to have originated from ASS about 70 million years ago (Nomiyama et al. 1986). Most of the pseudogenes are also found in the chimpanzee, including a Y-linked gene (Daiger et al. 1982), and in the mouse multiple sequences exist as well (quoted in T.S. Su et al. 1984).

5.11 Actin-like Pseudogene (ACTGP2)

The case of the actin-like sequence on the Y chromosome appears to be similar to that of the ASS pseudogene 6. Actin sequences form a multigene family dispersed over many chromosomes including X and Y, and some 50 of them have been detected (quoted in Koenig et al. 1985; see also HGM11). Many of these sequences represent processed pseudogenes, but at least six are expressed genes, coding for muscle and cytoskeletal actins (Heilig et al. 1984). Using a cDNA probe corresponding to α actin of human skeletal muscle (Hanauer et al. 1983), loci on the X and Y chromosomes were detected and assigned to Xp11–q11 and Yq11 by somatic cell hybrid analyses and deletion mapping (Heilig et al. 1984; Koenig et al. 1985). The Y-linked pseudogene has been designated ACTGP2 (McAlpine et al. 1988). The occurrence of actin-like sequences on both the X and Y chromosomes may be fortuitous and not due to the presence of conserved linkage groups, in particular since outside the actin-like sequences the homology ceases. In addition, in cercopithecoids, where no Y-linked actin locus has been detected, sequences homologous to the Y actin flanking sequences are autosomal (Koenig et al. 1985). However, in the chimpanzee, a male-specific DNA fragment has been identified with a human probe detecting a Y actin flanking sequence (Koenig et al. 1985).

5.12 Retroviral Sequences NP2 (RVNP2)

Screening of a human DNA library with a probe derived from a human endogenous retroviral sequence identified various clones, one of which, λNP2 was further studied because NP2-related sequences are highly polymorphic in man (Silver et al. 1987). Genomic Southern blots of male and female DNAs showed that NP2 is located on the Y chromosome, which also contains one other sequence closely related to NP2 retroviral and 3'-flanking sequences. Transcription activity was not detectable. It is assumed that this retroviral sequence was integrated randomly during evolution of the Y chromosome, and that this sequence was then duplicated. No functional significance can be ascribed to NP2.

6 Y-determined Phenotypic Characters

There are some genes postulated to exist on the Y chromosome by phenotypic mapping. Their assignment to the Y rests on male-to-male transmission of a character if compatible with fertility, and on deletion mapping in the case of sterility. Though individual gene loci in terms of DNA sequences have not yet been identified for these characters, current efforts will result in their isolation in the near future.

6.1 A Gene Controlling Spermatogenesis (AZF)

In a survey of subfertile males, Tiepolo and Zuffardi (1976) identified a number of azoospermic cases with a deletion of the distal region of Yq11. They postulated a gene controlling spermatogenesis located in this region. This gene was subsequently named AZF for *a*zoospermia *f*actor (Goodfellow et al. 1985b). Histological examination of cases with a deletion of distal Yq and azoospermia showed various degrees of germ cell degeneration (Chandley et al. 1986, 1989; Hartung et al. 1988). In EM microspreads of spermatocytes at pachytene of meiosis a proteinaceous body is visible on the Y axis in the more distal part of the nonpairing segment (Speed and Chandley 1990). As discussed by these authors, this proteinaceous body might be functionally related to the AZF gene, similar to the protein-RNA complexes formed on lampbrush chromosomes at the sites of fertility genes during meiosis of *Drosophila hydei* (Vogt and Henning 1986). Indeed, using a fertility gene sequence derived from *Drosophila hydei*, Vogt et al. (1991) identified a sequence family, pY6H, which maps to Y chromosome interval 6 of Vergnaud et al. (1986). The repetitive sequence composition of this locus is similar to that of *Drosophila* fertility genes, and the authors therefore consider it a candidate sequence for functional parts of the AZF locus. Other attempts to regionally map the putative AZF gene were performed by molecular analysis of Y-autosome translocations, confirming the location of the gene in deletion interval 6 on distal Yq11 (Andersson et al. 1988), and its location was further refined by a detailed study on the order of loci on Yq, resulting in its assignment to Yq11.23 (Bardoni et al. 1991). In addition, two unrelated sterile males with de novo microdeletions in Yq11.23 have been described (Vogt et al. 1992). The putative AZF gene may belong to an ancestral linkage group including TDF (Fig. 2) and presumably other genes involved in male sex determination and differentiation, because it appears to have a basic function in germ cell migration/proliferation during early embryonic life. In the mouse, a candidate spermatogenesis gene has recently been characterized which is located on the Y chromosome short arm in the so-called Sxr[b] region (Mitchell et al. 1991; Kay et al. 1991). This gene has a homologue on the X chromosome that is largely identical with the gene for the human ubiquitin-activating enzyme E1 (UBE1), which is also located on the X chromosome and escapes X-inactivation (Weissenbach and Goodfellow 1991). However, a homologous sequence has not been detected on the human Y chromosome.

6.2 Anti-Turner Genes

Turner's syndrome is characterized by gonadal dysgenesis and short stature and a number of variable featurs including pterygium colli, low hairline, lymphedema, and other abnormalities. Embryonic lethality is over 99% (Hassold 1986). So far, the genetic reason for Turner's syndrome and its high lethality are unexplained. Usually, this condition is associated with a 45,X karyotype; however, Turner characteristics are also observed in cases with deletions of segments of either the X or Y chromosome. It has been argued that the syndrome is the consequence of haploinsufficiency for a gene or genes present on both the X and Y chromosome and escaping inactivation (Ferguson-Smith 1965). These so far unknown genes on the Y chromosome should compensate in the male for the other X chromosome in the female, thus preventing Turner's syndrome, and therefore they can be named anti-Turner genes. Gonadal dysgenesis is not to be included among the disorders prevented by an anti-Turner gene, because it occurs also in XY females having a deficiency of TDF due to mutation or deletion (see Sect. 7, on sex determination). Thus, ovarian differentiation in the human female should depend on some as yet unknown gene(s) which are X specific and escape X-inactivation. It is the somatic anomalies in Turner's syndrome that should be due to the absence or functional impairment of genes shared by X and Y chromosomes.

An approach to identifying these genes is to analyze cases with Turner characteristics having small deletions in one of the sex chromosomes, and there is abundant literature about such cases. In XY females with Turner characteristics but normal height, terminal deletions were found in the Y chromosome short arm including the pseudoautosomal region and adjacent segments (e.g. Magenis et al. 1984a; Disteche et al. 1986b; Affara et al. 1987; Cantrell et al. 1989; Blagowidow et al. 1989; Levilliers et al. 1989). More refined deletion mapping assigned the critical segment to a subregion of interval 1 A (Fig. 3 B; Fisher et al. 1990b). RPS4Y located in this subregion was considered a candidate gene for Turner's syndrome, but this has become less likely in view of the gene expression studies discussed above under Sect. 5.4. In addition, it remains to be seen if the complex deletion situation in the case studied by Fisher et al. (see also Page et al. 1990b) definitely rules out other segments on Yp. Therefore, the hunt for anti-Turner genes is still at its beginning.

An attempt to map the gene putatively responsible for the short stature consistently occuring in Turner's syndrome was made by Henke et al. (1991). They studied two cases with partial monosomy of the pseudoautosomal region and localized this putative gene in a segment including two CpG islands and the CSF2R gene (see Sect. 5.1).

The putative haploinsufficiency causing Turner's syndrome obviously does not occur in the X0 condition of the mouse which is a fertile female. It is assumed that in this species the homologous "anti-Turner" genes undergo inactivation; therefore, no dosage difference would exist (Ashworth et al. 1991).

6.3 Growth Control, Y Chromosome-Influenced (GCY)

There is evidence for a gene or genes on the Y chromosome influencing growth in terms of body height and tooth size. Measurements of tooth size parameters in a man with most of the long arm of the Y chromosome deleted gave significantly smaller values than controls, while another man with only the distal part of the Y chromosome deleted had normal values (de la Chapelle and Alvesalo 1979; Alvesalo and de la Chapelle 1979, 1981). Based on these findings, a gene influencing growth was postulated to be located in region Yq11, possibly in its proximal part. This location coincides also with body height in cases with deletions of the Y chromosome, in particular XX males (see Sect. 7.1) who are significantly shorter than normal XY males. This putative gene has been named GCY for growth control of the Y chromosome, as introduced in HGM8 (McAlpine et al. 1985). These observations await confirmation and further refinement. The possibility is also discussed that GCY and AMGY may be one and the same gene (Nakahori et al. 1991a; Saido et al. 1992), and if this is the case, GCY should be located on proximal Yp.

6.4 Histocompatibility Y Antigen (H-Y)

Histocompatibility Y antigen is a minor histocompatibility antigen originally detected in female inbred mice which rejected isogenic male skin grafts (Eichwald and Silmser 1955). A serological assay was developed by Goldberg et al. (1971), and H-Y antigen can also be detected by cytotoxic T lymphocytes (Goldberg et al. 1973). In spite of various attempts to characterize H-Y antigen biochemically, so far only immunological definitions are applicable. There is evidence that more than one gene codes for proteins with H-Y activity, because the H-Y phenotype depends on the method used for its detection. Thus, three H-Y antigens may be defined, and following a proposal by Wiberg (1987), the antigen as detected by transplantation assays is termed, H-Yt, that identified by cytotoxic T lymphocytes H-Yc, and the serologically demonstrable one H-Ys. Because the term "H-Y" was originally coined to designate the transplantation antigen (Dillingham and Silvers 1960), it was proposed that the serological antigen be referred to as "SDM" for serologically detectable male antigen (Silvers et al. 1982). We consider this term misleading because if present in the female, this antigen should not be named "SDM". A thorough discussion of these antigens was provided by Wiberg (1987). H-Yt and H-Yc may be one and the same, and since mapping data with respect to the Y chromosome exist only for H-Yc and H-Ys, we will restrict the present discussion to these two putative genes.

H-Ys was found to be associated not only with the male sex in mammals, but also with the heterogametic sex in nonmammalian vertebrates (for review, see Wachtel 1983). Because of this association and its phylogenetic conservation, it was proposed to be responsible for male sex determination (Wachtel et al. 1975). Various pieces of evidence led to the proposal that the H-Ys structural gene is on an autosome while it is controlled by a gene on the Y chromosome; furthermore, it was proposed that H-Ys is not the gene for testis determination (TDF), which was assumed to be a major controlling gene, but rather a gene involved in primary gonadal differentiation and coding for a tissue hormone or morphogene (Wolf 1978, 1981, 1985). Expression of H-Ys in XX males (Yamada and Isurugi 1981; Wachtel 1983) carrying only the distal segment of the Y chromosome short arm favours the assumption that the postulated H-Ys controlling gene is in the sex-determining region of the Y. Correspondingly, phenotypic females with deletion of the short arm of the Y chromosome (and Turner's syndrome) proved to be negative for H-Ys (Rosenfeld et al. 1979; Kelly et al. 1984). So far, the possibility has not been ruled out that this controlling gene is TDF, because the H-Ys structural gene may be a gene functioning secondarily to TDF (Wolf 1988). Complementary to H-Ys, H-Yc has been found to be absent in XX males but present in females with a deletion of most of the short arm of the Y chromosome (E. Simpson et al. 1987). Therefore, H-Yc was assigned to Yp11.2–qter (Davies et al. 1987). Recently, again by deletion mapping, the location of H-Yc was narrowed down to a proximal portion of interval 6 of Vergnaud et al. (1986) (Cantrell et al. 1992).

In the mouse, H-Yc is not expressed in males carrying the Sxr[b] mutation (McLaren et al. 1984). This mutation causes a failure in spermatogenesis (Burgoyne et al. 1986), and H-Yc may therefore have some function in germ cell proliferation/differentiation (Burgoyne et al. 1986). However, in these mutant mice, H-Ys may be not expressed as well (Goldberg 1988), and if confirmed, this finding would argue against H-Ys being a morphogene responsible for primary testis differentiation. Interestingly, Müller and Wachtel (1991) found that H-Y antisera cross-react specifically with Müllerian inhibiting substance (MIS), thus opening up the possibility that H-Ys and MIS are one and the same. Indeed, H-Ys is secreted by testicular Sertoli cells (Zenzes et al. 1978a), as is the case with MIS. Furthermore, apart from its function of preventing the development of female ducts, MIS has been shown to sex-reverse fetal ovaries in vitro (Vigier et al. 1989), similar to H-Ys (Zenzes et al. 1978b; Müller and Urban 1981). There is an indication, however, that H-Ys antisera also cross-react with some other proteins (Lau et al. 1989b; H. Su et al. 1992).

6.5 Gonadoblastoma Locus on Y Chromosome (GBY)

Female patients carrying a Y chromosome or parts of it (XY gonadal dysgenesis; Turner's syndrome with XO/XY mosaicism or a cell line with a Y
fragment) have a predisposition for developing gonadoblastoma (Simpson
1976). Therefore, a gene on the Y chromosome was postulated to act as an
oncogene, if present in a dysgenetic gonad, and named GBY for *gonadoblastoma locus on Y* chromosome (Page 1987). This gene should have a
physiological function in normal males, presumably in the gonad. By deletion mapping in cases with XY gonadal dysgenesis and in Turner mosaics
with segments of the Y chromosome deleted, the short arm (Yp) was largely
excluded, and the putative gene should be located within Yp11.1–Yqter
(Page 1987; De Arce et al. 1992). So far, it cannot yet be excluded that GBY
is identical with one of the other genes postulated to be located within this
region on phenotypic grounds, i.e., H-Yc, GCY or AZF (Page 1987).

7 Sex Determination

The male-determining function of the Y chromosome became apparent
when, in the emerging field of human cytogenetics, a Klinefelter patient was
shown to have the sex chromosome constellation XXY (Jacobs and Strong
1959) and a Turner patient the constellation XO (Ford et al. 1959). Subsequently, subjects with various numbers of X chromosomes were found
who were male in the presence but female in the absence of a Y chromosome. Cytogenetic studies on structural aberrations indicated that the
short arm of the Y chromosome is crucial for male sex determination (for
review, see Davis 1981). A closer analysis of the sex-determining region,
however, became possible only after the introduction of DNA techniques.
A clue which finally led to the identification of the putative gene for the
testis-determining factor (TDF) came from the molecular study of XX
males and XY females with gonadal dysgenesis. This enigmatic discrepan-
cy between karyotype and phenotype prompted Ferguson-Smith, as early as
1966, to postulate that these conditions originate by X-Y interchange,
transferring the testis-determining region from the Y to the X, resulting in
XX^Y males, while the reciprocal product is a Y chromosome lacking this
region and carrying an X chromosomal segment in its place, thus resulting
in XY^X females (Fig. 5). This hypothesis proved to be correct, at least in a
number of cases. The molecular analysis of the Y chromosome segments
present in XX males and deleted in XY females allowed the definition of a
smallest region of overlap, and within this region a sequence necessary and
sufficient for male determination was eventually identified.

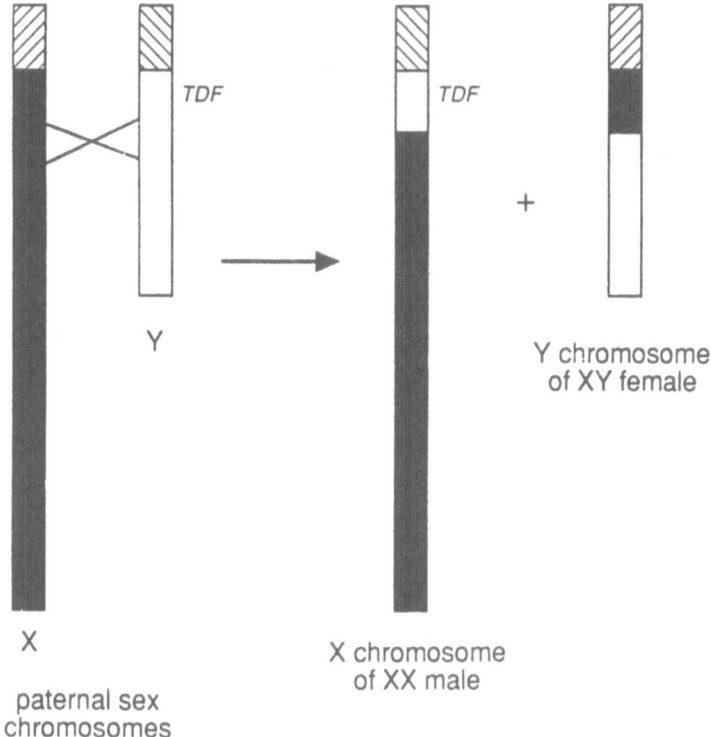

Fig. 5. Accidental terminal X-Y interchange in the etiology of XX males and XY females with gonadal dysgenesis. Aberrant X-Y recombination outside the pseudoautosomal pairing region (*hatched*) leads to transfer of TDF to the paternal X chromosome, giving rise to an XX male. The reciprocal product, a Y chromosome with TDF deleted, is found in some sex-reversed XY females with gonadal dysgenesis. The breakpoint on the X chromosome can be in the X-specific segment, as indicated, or within the pseudoautosomal region

7.1 XX Males

XX males are sterile, with small testes, azoospermia, and reduced levels of testosterone; otherwise, they are largely asymptomatic. The incidence is about 1 in 20 000 newborn boys (de la Chapelle 1972). They have an apparently normal 46,XX female karyotype. However, careful measurements on the X chromosomes of XX males revealed a significant size difference, the short arm of one X chromosome being slightly increased in around 70% of the cases studied (Evans et al. 1979). The authors interpreted these findings as evidence for X-Y interchange.

The first evidence for the presence of Y-specific DNA sequences in XX males was provided by Guellaen et al. (1984), using various hybridization probes derived from a Y chromosome library. The spectrum of positive signals for these probes varied among the cases studied, implying that the breakpoints on the Y chromosome were different. Thus, the presumed accidental X-Y interchange occurred at different locations in each case. These initial findings were subsequently confirmed in many other studies (Page et al. 1985; Vergnaud et al. 1986; U. Müller et al. 1986a; U. Müller 1987; Affara et al. 1986a, 1987; Buckle et al. 1987; Ferguson-Smith et al. 1987; Brøndum Nielsen et al. 1988; Schempp et al. 1989a). However, a fraction of the XX males analyzed proved to type negative for Y-specific DNA.

The X-Y interchange hypothesis of Ferguson-Smith (1966) was confirmed by DNA in situ hybridization; Y-specific DNA was indeed shown to be located at the distal short arm of one of the X chromosomes in metaphase chromosomes of XX males (Magenis et al. 1984b, 1987; Andersson et al. 1986). In a sample of 11 XX males, random X-inactivation was observed in ten cases, while in one male with a larger Xp segment including STS deleted, the Y-DNA-carrying X chromosome was preferentially inactivated (Schempp et al. 1989a). Furthermore, transfer of a terminal portion of Yp onto Xp in exchange for a terminal portion of Xp (Fig. 5), as predicted by Ferguson-Smith, was demonstrated by following the inheritance of pseudoautosomal alleles in XX male families (Page et al. 1987b; Petit et al. 1987). Inheritance of the entire pseudoautosomal region from the paternal Y chromosome and loss of all or the distal part of the pseudoautosomal region from the paternal X chromosome was found. In one case, this abnormal, terminal X-Y interchange resulted from homologous recombination between two Alu elements, one from distal Yp and one from within the X pseudoautosomal region, within otherwise nonhomologous regions (Rouyer et al. 1987).

The varying size of the Y segment transferred to one of the X chromosomes in XX males and some additional cases of Y deletions were used to construct a deletion map of the Y chromosome (Vergnaud et al. 1986; U. Müller et al. 1986a; Affara et al. 1987). The map constructed by Vergnaud et al. (1986) is the one most commonly used and comprises seven intervals defined by the presence or absence of Y-specific DNA segments in individual cases (see Fig. 3 A). Interval 1, defined by an XX male carrying the smallest Y segment in the sample studied, is also present in all other Y-positive XX males, and TDF must therefore lie in this interval. Later on, XX males with still smaller segments of the Y chromosome were found and interval 1 was subdivided (Page et al. 1987c; Palmer et al. 1989). By breakpoint analysis of such cases, the critical segment which should contain TDF was finally narrowed down to a 35-kb region proximal to the pseudoautosomal boundary, and subcloning of this region eventually resulted in the identification of SRY as a candidate sequence for TDF (Sinclair et al. 1990; see Sect. 7.4).

In addition to XX males, two other, less frequent classes of sterile males exist with karyotypes devoid of a Y chromosome. One class, the XXX (triplo-X) males, of which only three cases are known, are similar to XX males in having Yp material translocated onto one of their X chromosomes, but they have an additional X chromosome due to maternal or paternal nondisjunction (Annerén et al. 1987; U. Müller et al. 1987; Scherer et al. 1989b). The second class, the XO males, results from translocation of, in most cases, all of Yp including the Y centromere onto an autosome, leading to monosomy for the corresponding autosomal segment, which may result in congenital malformations. Following the initial molecular studies (Schempp et al. 1985; Maserati et al. 1986; Disteche et al. 1986a), over ten such cases have now been analyzed (listed in Hemel et al. 1992). The presence of interval 1 in all these individuals accounts for their male phenotype.

7.2 XY Females with Gonadal Dysgenesis

This condition, also named Swyer's syndrome, is found in sterile females with an apparently normal 46,XY male karyotype who are characterized by streak gonads, primary amenorrhea due to hypergonadotropic hypogonadism, poorly developed secondary sex characteristics, and predisposition to gonadal neoplasia (J.L. Simpson 1976; J.L. Simpson et al. 1981). While in a few cases deletions of the distal part of the Y chromosome are just detectable cytogenetically (Rosenfeld et al. 1979; Magenis et al. 1984a; Disteche et al. 1986b), in most cases such deletions become evident only by molecular analysis (Disteche et al. 1986b; U. Müller et al. 1986b; U. Müller 1987; Affara et al. 1987; Ferguson-Smith et al. 1987; Cantrell et al. 1989; Blagowidow et al. 1989; Levilliers et al. 1989). All such cases lack interval 1 of Vergnaud et al. (1986), probably due to an abnormal terminal X-Y interchange with transfer of Xp sequences to Yp, as actually shown by pedigree analysis and in situ hybridization in two cases (Levilliers et al. 1989). They thus represent the true mirror image of XX males as postulated by Ferguson-Smith (1966) (Fig. 5). It is intriguing that such Yp-deleted XY females seem to occur with a frequency significantly lower than that of Y-positive XX males. Only some 10%–20% of XY females studied in larger surveys (Ferguson-Smith et al. 1987; U. Müller 1987, Cantrell et al. 1989; our own unpublished data) have deletions on Yp, and because most such cases show Turner stigmata, a higher fetal mortality, as in Turner's syndrome, may account for the distortion in the ratio of X-Y interchange XY females to Y-positive XX males (Levilliers et al. 1989). After the discovery of SRY, an additional

10%–15% of XY gonadal dysgenesis females have revealed mutations within this gene (see below). However, the great majority of XY females with gonadal dysgenesis have a completely intact Y chromosome, pointing to the existence of a gene or genes acting secondarily to TDF which are not Y-linked and which must be affected by mutations in these cases. In fact, familial cases exist which indicate X-linked recessive or male-limited autosomal recessive, or dominant, inheritance (J.L. Simpson et al. 1981; Ostrer et al. 1989).

7.3 XX True Hermaphrodites·

True hermaphroditism is characterized by the simultaneous presence of testicular and ovarian tissues in the gonads. Some 60% of the cases have a normal female karyotype (van Niekerk and Retief 1981). The presence of testicular tissue has prompted several authors to search for Y-specific DNA in the genome of these subjects. Using various DNA probes from Yp and from the testis-determining region, no hybridization signal was obtained in the large majority of cases (Vergnaud et al. 1986; Waibel et al. 1987; Ramsay et al. 1988; Damiani et al. 1990; Abbas et al. 1990). Only a few exceptions were found to be positive for probes derived from the Y-pseudoautosomal boundary or the immediately adjacent region (Palmer et al. 1989; Jäger et al. 1990b; Nakagome et al. 1991). The cases studied by Palmer et al. were included in the search for TDF, resulting in the identification of the candidate sequence SRY (Sinclair et al. 1990).

A phenotypic analysis of Y-negative XX males indicates that they differ from classical, Y-positive XX males in exhibiting ambiguous genitalia (Ferguson-Smith et al. 1990; Abbas et al. 1990). It cannot be excluded that some or all of these cases are in fact true hermaphrodites, with some ovarian tissue undetected in their gonads. The observation that XX males can coexist with XX true hermaphrodites in the same pedigree (Skordis et al. 1987) points to a common origin. Similar to the majority of females with XY gonadal dysgenesis, a gene (or genes) secondary to TDF and not located on the X chromosome may be involved. It has been speculated that activating mutations in an X-linked locus subject to inactivation may account for both XX males and XX true hermaphrodites, with random X-inactivation in the developing gonads of the latter, and nonrandom inactivation in the former (Affara 1991).

7.4 The Testis-determining Gene

The idea that sex difference is under the control of a single master regulatory gene has been elaborated by Ohno (1979). The assumption of a single testis-determining gene, however, is by no means sufficient to explain male sexual differentiation (Mittwoch 1992), and indeed, the male differs from the female in a multiplicity of gene activities, both qualitatively (Y chromosome) and quantitatively (X chromosome). Nevertheless, the idea of a master regulatory gene is compatible with the view of a complex interaction of many genes, such as has been shown to be at work in the sex determination and differentiation pathways in the nematode *Caenorhabditis elegans* and in *Drosophila* (Hodgkin 1990).

An early approach to tracing candidate genes for male development was to look for male-specific proteins. In 1955, Eichwald and Silmser identified a male-specific histocompatibility antigen by transplantation experiments in the mouse, subsequently named H-Y antigen (Billingham and Silvers 1960). After it became evident that H-Y antigen is male specific not only in the mouse, but in all mammals studied, as well as in the heterogametic sex of nonmammalian vertebrates, this evolutionary conservation pointed to a fundamental function, and the hypothesis was put forward that it might be the product of the testis-determining gene of the mammalian Y chromosome (Wachtel et al. 1975; Ohno 1976). Since the occurrence of H-Y antigen was associated with the presence of testicular tissue in the gonads rather than with the presence of the Y chromosome (e.g., XX males and XX true hermaphrodites), Wolf (1978) postulated that it is not the H-Y structural gene, but a gene controlling H-Y expression which is located on the Y chromosome; a constitutive mutation could then explain expression of H-Y antigen in the absence of the Y chromosome. After the finding of a mutant male mouse with testes but lacking H-Y antigen (H-Yc, as discussed above; McLaren et al. 1984), this factor was definitely excluded as the testis determiner, and it has also become largely unpopular as a candidate in primary testis differentiation. In this context, however, we refer to the possible identity of H-Ys and MIS as discussed under Sect 6.4.

The identification of ZFY as a transcribed gene located in the testis-determining region of the Y chromosome (Page et al. 1987c) focused the interest on this gene as the best candidate for TDF. ZFY indeed fulfills several requirements postulated for the TDF including male specificity, evolutionary conservation, and its deduced function as a transcription factor. However, it soon became evident that it did not function as a TDF. It was considered irritating that ZFY has an almost identical homologue on the X chromosome, ZFX (U. Müller, quoted in Roberts 1988). In marsupials, sequences homologous to ZFY were found to be autosomal (Sinclair et al.

1988). In the mouse, the expression patterns of Zfy-1 and Zfy-2 (the mouse homologues of ZFY) were studied in embryonic male gonads. Zfy-2 transcripts were not detected, and Zfy-1 expression was shown to be associated with germ cells rather than with the supporting somatic cell lineage (Koopman et al. 1989). Finally, XX males and XX true hermaphrodites have been found lacking ZFY but still possessing Y-specific DNA derived from around the pseudoautosomal boundary (Palmer et al. 1989; Jäger et al. 1990b). For these reasons, ZFY was excluded from being TDF.

Further analysis of the XX male/intersex cases studied by Palmer et al. (1989) resulted in the definition of a 35-kb segment between the pseudoautosomal boundary and the point of X-Y interchange (Sinclair et al. 1990). By subcloning of DNA from the 35-kb segment, only one among some 50 probes was identified as fulfilling the following conditions: non-repetitive, Y specific, and present in all eutherian mammals tested. Sequence analysis revealed an open reading frame corresponding to a 204 amino acid protein with a stretch of 80 amino acids characteristic of a DNA-binding motif known as the HMG box, found in *high mobility group* proteins and in several transcription factors (Jantzen et al. 1990). The gene that includes this motif was designated SRY (Sinclair et al. 1990). It appears to have no intron and gives rise to an mRNA of about 1.0 kb in adult testis. The DNA-binding motif indicates that SRY codes for a protein controlling transcription, and this fits well with the expectation that TDF should be a master regulatory gene. In the meantime, an SRY homologue has been shown to be present on and specific for the marsupial Y chromosome as well (O'Brien and Graves 1991).

Further evidence for SRY being TDF (Tdy in the mouse) came from studies in the mouse. Gubbay et al. (1990) demonstrated that the murine equivalent Sry is present in Sxr[b], the smallest part of the mouse Y chromosome known to be sex determining (McLaren et al. 1984), and is deleted from a mutant Y chromosome that has lost male sex-determining function (Lovell-Badge and Robertson 1990). Sry expression was shown to be confined to somatic tissues of the developing male gonad, excluding germ cells; indeed, differentiation of the somatic architecture of the testis is known to be independent of germ cells (McLaren 1985). Furthermore, Sry expression is limited to the period when testis differentiation starts; transcripts were first detected on day 10.5 of embryonic development of the mouse, and at day 13.5 transcription was no longer detectable (Koopman et al. 1990).

That SRY is at least required for sex determination was demonstrated by the fact that 10%–15% of XY females with gonadal dysgenesis carry a mutation in SRY within the HMG box domain (Fig. 6). Of the 14 cases reported (see legend to Fig. 6 for references), eight have been shown to be de novo; three cases, V60L, I90M, and F109S, are familial mutations where the XY females share the same amino acid substitution with their fathers (Berta et al. 1990; Jäger et al. 1991; Vilain et al. 1992; Harley et al. 1992), while for an additional three cases, M78T, K106I, and R133W, it is not known whether or not the mutations are inherited. For the cases demonstrated to have de novo mutations in SRY, these mutations must be responsible for the sex reversal. Amino acid substitutions at highly conserved positions like I90M, G95R, K106I, and F109S may affect proper folding of the HMG box domain, while substitutions at less conserved positions like V60L, M64I, I68T, M78T, and R133W may identify residues involved in sequence-specific binding of SRY to its supposed DNA target.

Finally, the most convincing evidence that TDF (Tdy) has indeed been identified was provided by producing transgenic mice (Koopman et al. 1991b). When transferred into the nucleus of female mouse zygotes, a 14-kb DNA fragment including the mouse SRY sequence caused sex reversal in a number of cases, and the chromosomally female embryo grew up as a male (though sterile). Thus, the transferred fragment was shown to be not only necessary, but also sufficient for male sex determination. The only

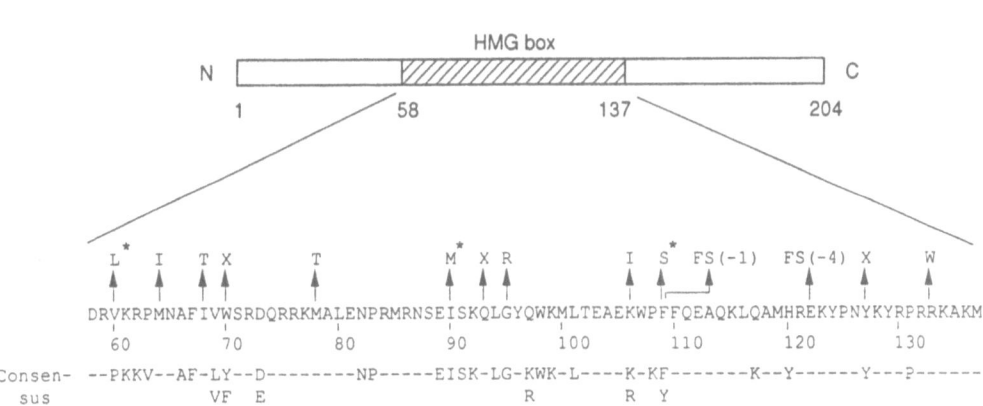

Fig. 6. Sex reversing mutations in SRY. All mutations identified in females with XY gonadal dysgenesis are located in the HMG box of SRY which spans residues 58–137 of the 204 amino acid residues of SRY. Single-letter code is used for the amino acid sequence. The mutations are from: V60L, M64I (Berta et al. 1990); E122FS(-4)(-4 frameshift) (Jäger et al. 1990a); F109S (Jäger et al. 1991); W70X, G95R (Hawkins et al. 1992), I90M, K106I (Harley et al. 1992); Q93X (McElreavey et al. 1992); F109FS(-1)(-1 frameshift) (J.R. Hawkins, G.D. Berkovitz and P.N. Goodfellow, personal communication); I68T, Y127X (K. McElreavey and M. Fellous, personal communication); M78T, R133W (N.A. Affara and M.A. Ferguson-Smith, personal communication). *X* denotes mutation to a stop codon. De novo mutations are not marked, inherited mutations are marked with an *asterisk*. M78T, K106I, and R133W may or may not be de novo mutations. The HMG box consensus sequence is modified from Harley et al. (1992)

transcript expressed from the 14-kb fragment if transfected into COS-1 cells is Sry mRNA, making it unlikely that another gene resides in this fragment (Walter et al. 1991).

In recent reports, the binding ability of normal and mutant SRY/Sry protein to particular DNA sequences has been studied, demonstrating that SRY/Sry is indeed a DNA-binding protein. Nasrin et al. (1991) have identified a rat protein, IRE-ABP, that binds to the insulin response element IRE-A of the glyceraldehyde-3-phosphate dehydrogenase gene. IRE-ABP contains a single HMG box that is 67% identical to that of mouse Sry and 98% identical to that of the mouse Sry-like protein a4 (Gubbay et al. 1990). Recombinant fusion proteins containing the HMG box region of Sry or IRE-ABP both bind with similar specificity to the IRE-A motif that contains the sequence 5' TTCAAAG 3', and introduction of the V60L or M64I mutations (Fig. 6) into the Sry fusion protein prevents this binding. Similarly, Harley et al. (1992) have expressed the complete human SRY protein and have shown that it binds to the sequence 5' AACAAAG 3' present in the CD3ε enhancer that is recognized by the T-lymphocyte transcription factor TCF-1, another HMG box protein (van de Wetering et al. 1991). The same sequence 5' AACAAAG 3' is also found upstream of both the human SRY and mouse Sry genes (cited in Harley et al. 1992) and is bound by SRY. Again, mutant SRY protein containing the amino acid substitutions V60L, M64I, G95R, or K106I (Fig. 6) showed no binding activity, while the familial I90M mutant showed greatly reduced DNA binding. The absence of any binding of the second familial mutant, V60L, observed in both studies might be explained by the potentially higher stringency of the in vitro assay compared to the in vivo situations (Harley et al. 1992). Finally, Giese et al. (1991) studied the binding properties of the TCF-1-related lymphoid-enhancer binding factor 1, LEF-1, also known as TCF-1α, to the T-cell antigen receptor (TCR) α enhancer that contains the sequence 5' TTCAAAG 3', identical to the IRE-A motif. They showed that a Y→S substitution in the single HMG box of LEF 1/TCF 1α that exactly corresponds in its position to the F109S familial SRY mutation (Fig. 6) leads to a ten fold reduction of binding to the TCR-α ehancer compared with the normal protein. The F109S mutant seems to be similar, therefore, to the other familial mutant I90M. The partial penetrance of the familial mutations may therefore be explained by stochastic fluctuation around a critical threshold value of a mutant SRY protein with reduced DNA binding; alternatively, the ability to functionally interact with different allelic forms of another, non-Y-encoded factor may be of decisive influence (Vilain et al. 1992).

It seems not very likely that IRE-A, CD3α, or TCRα are true targets for SRY. In vivo specificity for SRY might be brought about by interaction with another protein. For example, target gene specificity of muscle-specific transcription factors of the MyoD family seems to depend on such interactions with other transcription factors (Chakraborty and Olson 1991). The SRY protein might interact, however, with the sequence motif upstream of the SRY gene in vivo, autoregulating its own expression.

The spatiotemporal expression of SRY in the embryonic gonad requires that this gene itself is regulated. If SRY is TDF and as such the primary controlling factor for testis determination, the question arises as to what controls the controller. Such a controlling element cannot be Y linked, as shown by the Sry transgenic mice, and must be shared by both sexes. The respective gene should be active constitutively, regulating TDF (SRY) in the male sex, but possibly also other non-Y-linked genes.

The question remains which gene or genes are controlled by SRY if it codes for a transcription factor. It is to be expected that SRY is at the top of a cascade which finally results in testicular morphogenesis. In human beings, as in other mammals, some sex-reversal conditions are known showing an X-linked or autosomal mode of inheritance, e.g., some inherited forms of XY gonadal dysgenesis (J.L. Simpson et al. 1981), partial duplications of Xp (Bernstein et al. 1980; Scherer et al. 1989a), and campomelic dysplasia with sex reversal (Dagna Bricarelli et al. 1981). These conditions may involve genes acting secondarily to SRY. If these genes could be mapped and identified by positional cloning strategies, they might well serve as candidate loci for the action of SRY.

As outlined above, the evidence for equating SRY with TDF is compelling. One puzzle remaining is the SRY-positive, ZFY-negative cases. Of the four cases described by Palmer et al. (1989), only one is a classical XX male, two are similar to the Y-negative XX males with sexual ambiguities, and one is a true hermaphrodite, as are the two cases described by Jäger et al. (1990b) and Nakagome et al. (1991). If SRY is TDF, why are not all SRY-positive, ZFY-negative cases fully masculinized, as are (almost) all SRY-positive, ZFY-positive XX individuals? One explanation put forward by Burgoyne (1989) is that spreading of X-inactivation into the Y segment on Xp may affect the SRY locus in the SRY-positive, ZFY-negative cases, but not in the SRY-positive, ZFY-positive cases, where the interposition of more Y chromosome material buffers against this spreading. This explanation may hold as long as the translocated Yp material is at Xp22.3. If the translocation breakpoint on Xp is significantly more proximal, preferential inactivation of the Y-bearing X chromosome may occur, leading to true hermaphroditism even in an SRY-positive, ZFY-positive XX individual (Berkovitz et al. 1992); however, if such preferential inactivation occurs with the breakpoint proximal to the STS locus but still in band Xp22.3, a classical XX male with testes results (Schempp et al. 1989a). It therefore seems as if spreading of X-inactivation into the translocated Y segment and its effect on SRY expression may depend both on the size of the Y segment and on the position of the translocation breakpoint on Xp.

8 Conclusions

The human Y chromosome presents itself as a composite chromosome of recent origin. While convincing evidence confirms the hypothesis put forward by Ohno (1967) that the human sex chromosomes are derived from an original pair of autosomes, the sequences conserved from that ancestral past in the present-day Y chromosome appear to be confined to an essential minimum. It may be assumed that these sequences are restricted to genes involved in male development, including testis determination and male germ cell differentiation. These genes are not even directly linked any more, but are interrupted by sequences derived from other chromosomes, mainly from the X chromosome. Nevertheless, the requirement of several genes for male development must be the reason for the existence and the maintenance of a Y chromosome as such, thus guaranteeing the stability of sexual dimorphism.

Surprisingly, not even the homologous pairing segment between the X and Y chromosomes, the pseudoautosomal region, is a remainder of the ancestral homologous autosomal pair of chromosomes. Comparative studies of various mammalian species allow us to conclude that this region is of autosomal origin and was translocated to the X chromosome, and it may have been transferred to the Y chromosome by a meiotic recombination event. There are other segments of the Y chromosome showing homology with the X chromosome, and their different degrees of sequence divergence indicate that they were acquired at different times during evolution. Their present arrangement on the human Y chromosome is due in part to inversions, one of which, presumably a pericentric inversion, disrupted an originally much larger pseudoautosomal region and placed a part of it on the proximal long arm. With this event, the testis-determining region came into close proximity to the present pseudoautosomal region, thus giving rise to sex inversion by accidental X-Y interchange. X-Y homology outside the pseudoautosomal region is the reason for occasional pairing and crossing over during meiosis, resulting in X-Y translocations.

The Y chromosome is comprised up to 30%–40% of repetitive DNA which is not transcribed. The respective repeat elements may have invaded the Y chromosome by random events, and they generally share homologies of varying degree with repeats on other chromosomes. Their abundance on the Y chromosome is assumed to be a consequence of the genetic isolation of the differential part of this chromosome, so that amplification and possible subsequent deletion could take place without interfering with the requirements for pairing homology in meiosis. Due to the nature of this repetitive DNA, the Y chromosome exhibits great interindividual variation without phenotypic consequences. Part of this variation is reflected in the relatively high frequency of inversions, which again are explained by the meiotic isolation of the differential segment.

A number of genes and definitive or putative pseudogenes, including those postulated to exist on phenotypic grounds, are known to exist on the Y chromosome. Apart from genes in the pseudoautosomal region, which should be in a state similar to homologous autosomal genes, several genes in the differential segment originate from secondary transfer of larger DNA segments, and some of them have degenerated, or appear to be in the process of degeneration, to become pseudogenes. Sequence comparisons with the respective functional genes indicate that they themselves must have been functional until the recent past. Several of these genes and pseudogenes are located in the postulated ancestral pseudoautosomal region traced back to a pericentric inversion; consequently, their homologues are adjacent to the pseudoautosomal region of the X chromosome. Most or all of them escape X-inactivation at least partially, pointing to their pseudoautosomal origin, and this state presumably dates back to the time when the Y homologue was still a functional equivalent.

A number of genes have been shown or suggested to be involved in sex determination and male sexual differentiation; most of them are Y specific, as is to be expected. Among the Y-specific loci, SRY and AZF have been shown to play a role in sexual development, and this may also be true for TSPY and H-Y. GCY influences the male phenotype. For ZFY, a function in sex differentiation is also not excluded, though it has a homologue on the X chromosome. Thus, up to six genes on the Y chromosome, five of which are Y specific, have already been defined that are or may be involved in male development. A large number of Y-specific anonymous sequences remain, some of which may turn out to be transcribed, and their function has still to be defined. It is to be expected that on the Y chromosome a number of other genes await detection which may have a function in gonadal differentiation, in spermatogenesis, or in the prevention of the Turner phenotype.

The outstanding interest in the Y chromosome no doubt arises from its postulated sex-determining nature. Although the major testis-determining gene, TDF, has very likely been identified with SRY, sex determination is still far from being understood. We do not know how SRY is regulated and what its targets are, and its complex interactions will move research interest to genes on other chromosomes. However, its unique structure, its small size in terms of nonrepetitive DNA, and the relative paucity of genes makes it likely that the Y chromosome will be among the first human chromosomes whose genetic contributions are comprehensively understood.

Acknowledgements. We thank N.A. Affara, G.D. Berkovitz, M. Fellous, M.A. Ferguson-Smith, P.N. Goodfellow, J.R. Hawkins, and K. McElreavey for information on SRY mutations prior to publication. Helpful comments on the manuscript were provided by U. Müller, G. Rappold, A. Stewart, and M. Fraccaro. We are also indebted to Elke Kunstmann for her secretarial cooperation.

References

Abbas NE, Toublanc JE, Boucekkine C, Toublanc M, Affara NA, Job J-C, Fellous M (1990) A possible common origin of "Y-negative" human XX males and XX true hermaphrodites. Hum Genet 84:356–360

Affara NA (1991) Sex and the single Y. Bioessays 13:475–478

Affara NA, Ferguson-Smith MA, Tolmie J, Kwok K, Mitchell M, Jamieson D, Cooke A, Florentin L (1986a) Variable transfer of Y-specific sequences in XX males. Nucleic Acids Res 14:5375–5387

Affara NA, Florentin L, Morris N, Kwok K, Mitchell M, Cook A, Jamieson D, Glasgow L, Meredith L, Boyd E, Ferguson-Smith MA (1986b) Regional assignment of Y-linked DNA probes by deletion mapping and their homology with X-chromosome and autosomal sequences. Nucleic Acids Res 14:5353–5373

Affara NA, Ferguson-Smith MA, Magenis RE, Tolmie JL, Boyd E, Cooke A, Jamieson D, Kwok K, Mitchell M, Snadden L (1987) Mapping the testis determinants by an analysis of Y-specific sequences in males with apparent XX and XO karyotypes and females with XY karyotypes. Nucleic Acids Res 15:7325–7342

Affara NA, Chambers D, O'Brien J, Habeebu SSM, Kalaitsidaki M, Bishop CE, Ferguson-Smith MA (1989) Evidence for distinguishable transcripts of the putative testis-determining gene (ZFY) and mapping of homologous cDNA sequences to chromosomes X, Y and 9. Nucleic Acids Res 17:2987–2999

Alvesalo L, de la Chapelle A (1979) Permanent tooth sizes in 46,XX males. Ann Hum Genet 43:97–102

Alvesalo L, de la Chapelle A (1981) Tooth sizes in two males with deletions of the long arm of the Y-chromosome. Ann Hum Genet 45:49–54

Andersson M, Page DC, de la Chapelle A (1986) Chromosome Y-specific DNA is transferred to the short arm of X chromosome in human XX males. Science 233:786–788

Andersson M, Page DC, Pettay D, Subrt I, Turleau C, de Grouchy J, de la Chapelle A (1988) Y; autosome translocations and mosaicism in the aetiology of 45,X maleness: assignment of fertility factor to distal Yq11. Hum Genet 79:2–7

Annerén G, Andersson M, Page DC, Brown LG, Berg M, Läckgren G, Gustavson K-H, de la Chapelle A (1987) An XXX male resulting from paternal X-Y interchange and maternal X-X nondisjunction. Am J Hum Genet 41:594–604

Arnemann J, Epplen JT, Cooke HJ, Sauermann U, Engel W, Schmidtke J (1987) A human Y-chromosomal DNA sequence expressed in testicular tissue. Nucleic Acids Res 15: 8713–8724

Arnemann J, Jakubiczka S, Thüring S, Schmidtke J (1991) Cloning and sequence analysis of a human Y-chromosome-derived, testicular cDNA, TSPY. Genomics 11:108–114

Arrighi FE, Hsu TC (1971) Localization of heterochromatin in human chromosomes. Cytogenetics 10:81–86

Arveiler B, Vincent A, Mangel J-L (1989) Toward a physical map of the Xq28 region in man: linking color vision, G6PD, and coagulation factor VIII genes to an X-Y homology region. Genomics 4:460–471

Ashworth A, Rastan S, Lovell-Badge R, Kay G (1991) X-chromosome inactivation may explain the difference in viability of X0 humans and mice. Nature 351:406–408

Ballabio A, Carrozzo G, Gil A, Gillard B, Affara N, Ferguson-Smith MA, Fraser N, Craig I, Rocchi M, Romeo G, Andria G (1989) Molecular characterization of human X/Y translocations suggests their aetiology through aberrant exchange between homologous sequences on Xp and Yp. Ann Hum Genet 53:9–14

Bardoni B, Zuffardi O, Guioli S, Ballabio A, Simi P, Cavalli P, Grimoldi MG, Fraccaro M, Camerino G (1991) A deletion map of the human Yq11 region: implications for the evolution of the Y chromosome and tentative mapping of a locus involved in spermatogenesis. Genomics 11:443–451

Beaudet AL, Su TS, O'Brien WE (1982) Dispersion of argininosuccinate-synthetase-like human genes to multiple autosomes and the X chromosome. Cell 30:287–293

Beçak W, Beçak ML, Nazareth HRS, Ohno S (1964) Close karyological kinship between the reptilian suborder Serpentes and the class Aves. Chromosoma 15:606–617

Berkovitz GD, Fechner PY, Marcantonio SM, Bland G, Stetten G, Goodfellow PN, Smith KD, Migeon JC (1992) The role of the sex-determining region of the Y chromosome (SRY) in the etiology of 46,XX true hermaphroditism. Hum Genet 88:411–416

Bernstein R, Jenkins T, Dawson B, Wagner J, Dewald G, Koo GC, Wachtel SS (1980) Female phenotype and multiple abnormalities in sibs with a Y chromosome and partial X chromosome duplication: H-Y antigen and Xg blood group findings. J Med Genet 17: 291–300

Berta P, Hawkins JR, Sinclair AH, Taylor A, Griffiths BL, Goodfellow PN, Fellous M (1990) Genetic evidence equating SRY and the testis-determining factor. Nature 348: 448–450

Bickmore WA, Cooke HJ (1987) Evolution of homologous sequences of the human X and Y chromosomes, outside of the meiotic pairing segment. Nucleic Acids Res 15: 6261–6271

Billingham RE, Silvers WK (1960) Studies on tolerance of the Y chromosome antigen in mice. J Immunol 85:14–26

Blagowidow N, Page DC, Huff D, Mennuti MT (1989) Ullrich-Turner syndrome in an XY female fetus with deletion of the sex-determining portion of the Y chromosome. Am J Med Genet 34:159–162

Bobrow M, Pearson PL, Pike MC, El-Alfi OS (1971) Length variation in the quinacrine-binding segment of human Y chromosomes of different sizes. Cytogenetics 10:190–198

Borgaonkar DS, Hollander DH (1971) Quinacrine fluorescence of the human Y chromosome. Nature 230:52

Bostock CJ, Gosden JR, Mitchell AR (1978) Localisation of a male-specific DNA fragment to a sub-region of the human Y chromosome. Nature 272:324–328

Brøndum Nielsen K, Schwartz M, Sardemann H (1988) Investigation of three XX males by cytogenetic and DNA analyses. Hum Genet 78:179–182

Brown CJ, Ballabio A, Rupert JL, Lafreniere RG, Grompe M, Tonlorenzi R, Willard HF (1991a) A gene from the region of the human X inactivation centre is expressed exclusively from the inactive X chromosome. Nature 349:38–44

Brown CJ, Lafreniere RG, Powers VE, Sebastio G, Ballabio A, Pettigrew AL, Ledbetter DH, Levy E, Craig IW, Willard HF (1991b) Localization of the X inactivation centre on the human X chromosome in Xq13. Nature 349:82–84

Brown WRA (1988) A physical map of the human pseudoautosomal region. EMBO J 7: 2377–2385

Buckle V, Mondello C, Darling S, Craig IW, Goodfellow PN (1985) Homologous expressed genes in the human sex chromosome pairing region. Nature 317:739–741

Buckle VJ, Boyd Y, Fraser N, Goodfellow PN, Goodfellow PJ, Wolfe J, Craig IW (1987) Localisation of Y chromosome sequences in normal and 'XX' males. J Med Genet 24: 197–203

Burgoyne PS (1982) Genetic homology and crossing over in the X and Y chromosomes of mammals. Hum Genet 61:85–90

Burgoyne PS (1989) Thumbs down for zinc finger? Nature 342:860–862

Burgoyne PS, Levy ER, McLaren A (1986) Spermatogenic failure in male mice lacking H-Y antigen. Nature 320:170–172

Burk RD, Szabo P, O'Brien S, Nash WG, Yu L, Smith KD (1985) Organization and chromosomal specificity of autosomal homologs of human Y chromosome repeated DNA. Chromosoma 92:225–233

Cantrell MA, Bicknell JN, Pagon RA, Page DC, Walker DC, Saal HM, Zinn AB, Disteche CM (1989) Molecular analysis of 46,XY females and regional assignment of a new Y-chromosome-specific probe. Hum Genet 83:88–92

Cantrell MA, Bogan JS, Simpson E, Bicknell JN, Goulmy E et al. (1992) Deletion mapping of H-Y antigen to the long arm of the human Y chromosome. Genomics 13:1255–1260

Carritt B, Povey S (1979) Regional assignments of the loci AK_3, $ACON_s$, and ASS on human chromosome 9. Cytogenet Cell Genet 23:171–181

Chakraborty T, Olson EN (1991) Domains outside of the DNA-binding domain impart target gene specificity to myogenin and MRF4. Mol Cell Biol 11:6103–6108

Chandley AC, Goetz P, Hargreave TB, Joseph AM, Speed RM (1984) On the nature and extent of XY pairing at meiotic prophase in man. Cytogenet Cell Genet 38:241–247

Chandley AC, Ambros P, McBeath S, Hargreave TB, Kilanowski F, Spowart G (1986) Short arm dicentric Y chromosome with associated statural defects in a sterile man. Hum Genet 73:350–353

Chandley AC, Gosden JR, Hargreave TB, Spowart G, Speed RM, McBeath S (1989) Deleted Yq in the sterile son of a man with a satellited Y chromosome (Yqs). J Med Genet 26:145–153

Cohen MM, Shaw MW, MacCluer JW (1966) Racial differences in the length of the human Y chromosome. Cytogenetics 5:34–52

Cooke HJ (1976) Repeated sequence specific to human males. Nature 262:182–186

Cooke HJ, Schmidtke J, Gosden JR (1982) Characterisation of a human Y chromosome repeated sequence and related sequences in higher primates. Chromosoma 87:491–502

Cooke HJ, Fantes J, Green D (1983) Structure and evolution of human Y chromosome DNA. Differentiation 23:S48–S55

Cooke HJ, Brown WAR, Rappold GA (1984) Closely related sequences on human X and Y chromosomes outside the pairing region. Nature 311:259–261

Cooke HJ, Brown WRA, Rappold GA (1985) Hypervariable telomeric sequences from the human sex chromosomes are pseudoautosomal. Nature 317:687–692

Dagna Bricarelli F, Fraccaro M, Lindsten J, Müller U, Baggio P, Doria Lamba Carbone L, Hjerpe A, Lindgren F, Mayerová A, Ringertz H, Ritzén EM, Rovetta DC, Siccheio C, Wolf U (1981) Sex-reversed XY females with campomelic dysplasia are H-Y negative. Hum Genet 57:15–22

Daiger SP, Wildin RS, Su TS (1982) Sequences on the human Y chromosome homologous to the autosomal gene for argininosuccinate synthetase. Nature 298:682–684

Damiani D, Billerbeck AEC, Goldberg ACK, Setian N, Fellous M, Kalil J (1990) Investigation of the ZFY gene in XX true hermaphroditism and Swyer syndrome. Hum Genet 85: 85–88

Darling SM, Banting GS, Pym B, Wolfe J, Goodfellow PN (1986) Cloning an expressed gene shared by the human sex chromosomes. Proc Natl Acad Sci USA 83:135–139

Davies KE, Mandel JL, Weissenbach J, Fellous M (1987) Report of the committee on the genetic constitution of the X and Y chromosomes. Cytogenet Cell Genet 46:277–315

Davis RM (1981) Localisation of male-determining factors in man: a thorough review of structural anomalies of the Y chromosome. J Med Genet 18:161–195

De Arce MA, Costigan C, Gosden JR, Lawler M, Humphries P (1992) Further evidence consistent with Yqh as an indicator of risk of gonadal blastoma in Y-bearing mosaic Turner syndrome. Clin Genet 41:28–32

de la Chapelle A (1972) Nature and origin of males with XX sex chromosomes. Am J Hum Genet 24:71–105

de la Chapelle A, Alvesalo L (1979) Mapping of the growth promoting gene(s) on the human Y chromosome. Cytogenet Cell Genet 25:146–147

Disteche CM, Brown L, Saal H, Friedman C, Thuline HC, Hoar DI, Pagon RA, Page DC (1986a) Molecular detection of a translocation (Y;15) in a 45,X male. Hum Genet 74: 372–377

Disteche CM, Casanova M, Saal H, Friedman C, Sybert V, Graham J, Thuline H, Page DC, Fellous M (1986b) Small deletions of the short arm of the Y chromosome in 46,XY females. Proc Natl Acad Sci USA 83:7841–7844

Eichwald EJ, Silmser CR (1955) Communication. Transplant Bull 2:148–149

Ellis NA, Goodfellow PN (1989) The mammalian pseudoautosomal region. Trends Genet 5:406–410

Ellis NA, Goodfellow PJ, Pym B, Smith M, Palmer M, Frischauf A-M, Goodfellow PN (1989) The pseudoautosomal boundary in man is defined by an *Alu* repeat sequence inserted on the Y chromosome. Nature 337:81–84

Ellis NA, Taylor A, Bengtsson BO, Kidd J, Rogers J, Goodfellow P (1990a) Population structure of the human peusodoautosomal boundary. Nature 344:663–665

Ellis NA, Yen P, Neiswanger K, Shapiro LJ, Goodfellow PN (1990b) Evolution of the pseudoautosomal boundary in old world monkeys and great apes. Cell 63:977–986

Evans HJ, Buckton KE, Spowart G, Carothers AD (1979) Heteromorphic X chromosomes in 46,XX males: evidence for the involvement of X-Y interchange. Hum Genet 49:11–31

Ferguson-Smith MA (1965) Karyotype-phenotype correlations in gonadal dysgenesis and their bearing on the pathogenesis of malformations. J Med Genet 2:142–155

Ferguson-Smith MA (1966) X-Y chromosomal interchange in the aetiology of true hermaphroditism and of XX Klinefelter's syndrome. Lancet II:475–476

Ferguson-Smith MA, Affara NA, Magenis RE (1987) Ordering of Y-specific sequences by deletion mapping and analysis of X-Y interchange males and females. Development 101 [Suppl]:41–50

Ferguson-Smith MA, Cooke A, Affara NA, Boyd E, Tolmie JL (1990) Genotype-phenotype correlations in XX males and their bearing on current theories of sex determination. Hum Genet 84:198–202

Fisher EMC, Alitalo T, Luoh S-W, de la Chapelle A, Page DC (1990a) Human sex-chromosome-specific repeats within a region of peusodoautosomal/Yq homology. Genomics 7:625–628

Fisher EMC, Beer-Romero P, Brown LG, Ridley A, McNeil JA, Bentley Lawrence J, Willard HF, Bieber FR, Page DC (1990b) Homologous ribosomal protein genes on the human X and Y chromosomes: escape from X inactivation and possible implications for Turner syndrome. Cell 63:1205–1218

Ford CE, Jones KW, Polani PE, de Almeida JC, Briggs JH (1959) A sex chromosome anomaly in a case of gonadal dysgenesis (Turner's syndrome). Lancet I:711–713

Franco B, Guioli S, Pragliola A, Incerti B, Bardoni B et al. (1991) A gene deleted in Kallmann's syndrome shares homology with neural cell adhesion and axonal path-finding molecules. Nature 353:529–536

Fraser N, Ballabio A, Zollo M, Persico G, Craig I (1987) Identification of incomplete coding sequences for steroid sulphatase on the human Y chromosome: evidence for an ancestral peusodoautosomal gene? Development 101 [Suppl]:127–132

Freytag SO, Bock H-GO, Beaudet AL, O'Brien WE (1984) Molecular structures of human argininosuccinate synthetase pseudogenes. J Biol Chem 259:3160–3166

Frommer M, Prosser J, Vincent PC (1984) Human satellite I sequences include a male-specific 2.47-kb tandemly repeated unit containing one Alu family member per repeat. Nucleic Acids Res 12:2887–2900

Gearing DP, King JA, Gouth NM, Nicola NA (1989) Expression cloning of a receptor for human granulocyte-macrophage colony-stimulating factor. EMBO J 8:3667–3676

Geldwerth D, Bishop C, Guellaen G, Koenig M, Vergnaud G, Mandel J-L, Weissenbach J (1985) Extensive DNA sequence homologies between the human Y and the long arm of the X chromosome. EMBO J 4:1739–1743

Geller RL, Shapiro LJ, Mohandas TK (1986) Fine mapping of the distal short arm of the human X chromosome using X/Y translocations. Am J Hum Genet 38:884–890

Giese K, Amsterdam A, Grosschedl R (1991) DNA-binding properties of the HMG domain of the lymphoid-specific transcriptional regulator LEF-1. Genes Dev 5:2567–2578

Goldberg EH (1988) H-Y antigen and sex determination. Philos Trans R Soc Lond [Biol] 322:73–81

Goldberg EH, Boyse EA, Bennett D, Scheid M, Carswell EA (1971) Serological demonstration of H-Y (male) antigen on mouse sperm. Nature 232:478–480

Goldberg EH, Shen F-W, Tokuda S (1973) Detection of H-Y (male) antigen on mouse lymph node cells by the cell-to-cell cytotoxicity test. Transplantation 15:334–336

Goodfellow P (1983) Expression of the 12E7 antigen is controlled independently by genes on the human X and Y chromosomes. Differentiation 23:S35–S39

Goodfellow P, Banting G, Sheer D, Ropers HH, Caine A, Ferguson-Smith MA, Povey S, Voss R (1983) Genetic evidence that a Y-linked gene in man is homologous to a gene on the X chromosome. Nature 302:346–349

Goodfellow P, Darling S, Wolfe J (1985a) The human Y chromosome. J Med Genet 22: 329–344

Goodfellow PN, Davies KE, Ropers HH (1985b) Report of the committee on the genetic constitution of the X and Y chromosomes. Cytogenet Cell Genet 40:296–352

Goodfellow PJ, Darling SM, Thomas NS, Goodfellow PN (1986) A peusodoautosomal gene in man. Science 234:740–743

Goodfellow PN, Craig IW, Smith JC, Wolfe J (eds) (1987a) The mammalian Y chromosome: molecular search for the sex-determining factor. Development 101 [Suppl]:1–203

Goodfellow PJ, Darling S, Banting G, Pym B, Mondello C, Goodfellow PN (1987b) Pseudoautosomal genes in man. Development 101 [Suppl]:119–125

Gough NM, Gearing DP, Nicola NA, Baker E, Pritchard M, Callen DF, Sutherland GR (1990) Localization of the human GM-CSF receptor gene to the X-Y peusodoautosomal region. Nature 345:734–736

Graves JAM, Watson JM (1991) Mammalian sex chromosomes: evolution of organisation and function. Chromosoma 101:63–68

Gubbay J, Collignon J, Koopman P, Capel B, Economou A, Münsterberg A, Vivian N, Goodfellow P, Lovell-Badge R (1990) A gene mapping to the sex-determining region of the mouse Y chromosome is a member of a novel family of embryonically expressed genes. Nature 346:245–250

Guellaen G, Casanova M, Bishop C, Geldwerth D, Andre G, Fellous M, Weissenbach J (1984) Human XX males with Y single-copy DNA fragments. Nature 307:172–173

Hanauer A, Levin M, Heilig R, Daegelen D, Kahn A, Mandel JL (1983) Isolation and characterization of cDNA clones for human skeletal muscle α actin. Nucleic Acids Res 11: 3503–3516

Harbers K, Soriano P, Müller U, Jaenisch R (1986) High frequency of unequal recombination in peusodoautosomal region shown by proviral insertion in transgenic mouse. Nature 324:682–685

Harley VR, Jackson DJ, Hextall PJ, Hawkins JR, Berkovitz GD, Sockanathan S, Lovell-Badge R, Goodfellow PN (1992) DNA-binding activity of recombinant SRY from normal males and XY females. Science 255:453–456

Hartung M, Devictor M, Codaccioni JL, Stahl A (1988) Yq deletion and failure of spermatogenesis. Ann Genet 31:21–26

Hassold TJ (1986) Chromosome abnormalities in human reproductive wastage. Trends Genet 2:105–110

Hawkins JR, Taylor A, Berta P, Levilliers J, Auwera B van der, Goodfellow PN (1992) Mutational analysis of *SRY*: nonsense and missense mutations in XY sex reversal. Hum Genet 88:471–474

Hayman DL, Martin PG (1974) Mammalia I: monotremata and marsupialia. Chordata 4. In: John B (ed) Animal Cytogenetics. Borntraeger, Berlin

Heilig R, Hanauer A, Grzeschik KH, Hors-Cayla MC, Mandel JL (1984) Actin-like sequences are present on human X and Y chromosomes. EMBO J 8:1803–1807

Hemel JO van, Eussen B, Wesby-van Swaay E, Oostra BA (1992) Molecular detection of a translocation (Y;11)(q11.2;q24) in a newborn with signs of Jacobsen syndrome. Hum Genet 88:661–667

Henke A, Wapenaar M, Ommen G-J van, Maraschio P, Camerino G, Rappold G (1991) Deletions within the peusodoautosomal region help map three new markers and indicate a possible role of this region in linear growth. Am J Hum Genet 49:811–819

HGM1 (1974) Human Gene Mapping 1. New Haven Conference (1973). Cytogenet Cell Genet 13:1–216

HGM2 (1975) Human Gene Mapping 2. Rotterdam Conference (1974). Cytogenet Cell Genet 14:161–480

HGM3 (1976) Human Gene Mapping 3. Baltimore Conference (1975). Cytogenet Cell Genet 16:1–452

HGM9 (1987) Human Gene Mapping 9. Paris Conference (1987). Cytogenet Cell Genet 46:1–762

HGM10 (1989) Human Gene Mapping 10. New Haven Conference (1989). Cytogenet Cell Genet 51:1–1147

HGM11 (1991) Human Gene Mapping 11. London Conference (1991). Cytogenet Cell Genet 58:1–2197

Hodgkin J (1990) Sex determination compared in *Drosophila* and *Caenorhabditis*. Nature 344:721–728

Hultén M (1974) Chiasma distribution at diakinesis in the normal human male. Hereditas 76:55–78

Jacobs PA, Strong JA (1959) A case of human intersexuality having a possible XXY sex-determining mechanism. Nature 183:302–303

Jäger RJ, Anvret M, Hall K, Scherer G (1990a) A human XY female with a frame shift mutation in the candidate testis-determining gene *SRY*. Nature 348:452–454

Jäger RJ, Ebensperger C, Fraccaro M, Scherer G (1990b) A ZFY-negative 46,XX true hermaphrodite is positive for the Y peusodoautosomal boundary. Hum Genet 85:666–668

Jäger RJ, Pfeiffer RA, Scherer G (1991) A familial amino acid substitution in SRY can lead to conditional XY sex inversion. Am J Hum Genet 49:S219

Jantzen H-M, Admon A, Bell SP, Tjian R (1990) Nucleolar transcription factor hUBF contains a DNA-binding motif with homology to HMG proteins. Nature 344:830–836

Johnson CL, Charmley P, Yen PH, Shapiro LJ (1991) A multipoint linkage map of the distal short arm of the human X chromosome. Am J Hum Genet 49:261–266

Jones KW, Singh L (1985) Snakes and the evolution of sex chromosomes. Trends Genet 1:55–61

Just W, Geerkens C, Held KR, Vogel W (1992) Expression of RPS4X in fibroblasts from patients with structural aberrations of the X chromosome. Hum Genet 89:240–242

Kay GF, Ashworth A, Penny GD, Dunlop M, Swift S, Brockdorff N, Rastan S (1991) A candidate spermatogenesis gene on the mouse Y chromosome is homologous to ubiquitin-activating enzyme E1. Nature 354:486–489

Keitges E, Rivest M, Siniscalco M, Gartler SM (1985) X-linkage of steroid sulphatase in the mouse is evidence for a functional Y-linked allele. Nature 315:226–227

Keitges EA, Schorderet DF, Gartler SM (1987) Linkage of the steroid sulfatase gene to the sex-reversed mutation in the mouse. Genetics 116:465–468

Kelly TE, Wachtel SS, Cahill L, Barnabei VM, Willson-Suddath K, Wyandt HE (1984) X;Y translocation in a female with streak gonads, H-Y phenotype, and some features of Turner's syndrome. Cytogenet Cell Genet 38:122–126

Koenig M, Moisan JP, Heilig R, Mandel JL (1985) Homologies between X and Y chromosomes detected by DNA probes: localisation and evolution. Nucleic Acids Res 13: 5485–5501

Koller PC, Darlington CD (1934) The genetical and mechanical properties of the sex chromosomes: 1. *Rattus norvegicus.* J Genet 29:159–173

Koopman P, Gubbay J, Collignon J, Lovell-Badge R (1989) Zfy gene expression patterns are not compatible with a primary role in mouse sex determination. Nature 342:940–942

Koopman P, Münsterberg A, Capel B, Vivian N, Lovell-Badge R (1990) Expression of a candidate sex-determining gene during mouse testis differentiation. Nature 348:450–452

Koopman P, Ashworth A, Lovell-Badge R (1991a) The ZFY gene family in humans and mice. Trends Genet 4:132–136

Koopman P, Gubbay J, Vivian N, Goodfellow P, Lovell-Badge R (1991b) Male development of chromosomally female mice transgenic for *Sry.* Nature 351:117–121

Kunkel LM, Tantravahi U, Kurnit DM, Eisenhard M, Bruns GP, Latt SA (1983) Identification and isolation of transcribed human X chromosome DNA sequences. Nucleic Acids Res 11:7961–7979

Lau EC, Mohandas TK, Shapiro LJ, Slavkin HC, Snead ML (1989a) Human and mouse amelogenin gene loci are on the sex chromosomes. Genomics 4:162–168

Lau Y-F (1985) Organization of the human Y-specific *Hae*III 3.4-kb repeat sequences and their application in clinical diagnosis. In: Sandberg AA (ed) The Y chromosome: A. Basic characteristics of the Y chromosome. Liss, New York, pp 177–192

Lau Y-F, Chan K (1989) The putative testis-determining factor and related genes are expressed as discrete-sized transcripts in adult gonadal and somatic tissues. Am J Hum Genet 45:942–952

Lau Y-F, Chan K, Sparkes R (1989b) Male-enhanced antigen gene is phylogenetically conserved and expressed at late stages of spermatogenesis. Proc Natl Acad Sci USA 86: 8462–8466

Legouis R, Hardelin J-P, Levillier J, Claverie J-M, Compain S et al. (1991). The candidate gene for the X-linked Kallmann syndrome encodes a protein related to adhesion molecules. Cell 67:423–435

Levilliers J, Quack B, Weissenbach J, Petit C(1989) Exchange of terminal portions of X- and Y-chromosomal short arms in human XY females. Proc Natl Acad Sci USA 86: 2296–2300

Levy R, Dilley J, Fox RI, Warnke R (1979) A human thymus-leukemia antigen defined by hybridoma monoclonal antibodies. Proc Natl Acad Sci USA 76:6552–6556

Lovell-Badge R, Robertson E (1990) XY-female mice resulting from a heritable mutation in the primary testis-determining gene, Tdy. Development 109:635–646

Magenis RE, Tochen ML, Holahan KP, Carey T, Allen L, Brown MG (1984a) Turner syndrome resulting from partial deletion of Y chromosome short am: localization of male determinants. J Pediatr 105:916–919

Magenis RE, Tomar D, Sheehy R, Fellous M, Bishop C, Casanova M (1984b) Y short arm material translocated to distal X short arm in XX males: evidence from in situ hybridisation of a Y-specific single-copy DNA probe. Am J Hum Genet 36:102S

Magenis RE, Casanova M, Fellous M, Olson S, Sheehy R (1987) Further cytologic evidence for Xp-Yp translocation in XX males using in situ hybridization with Y-derived probe. Hum Genet 75:228–233

Mardon G, Mosher R, Disteche CM, Nishioka Y, McLaren A, Page DC (1989) Duplication, deletion, and polymorphism in the sex-determining region of the mouse Y chromosome. Science 243:78–80

Maserati E, Waibel F, Weber B, Fraccaro M, Gal A, Pasquali F, Schempp W, Scherer G, Vaccaro R, Weissenbach J, Wolf U (1986) A 45,X male with a Yp/18 translocation. Hum Genet 74:126–132

McAlpine PJ, Shows TB, Miller RL, Pakstis AJ (1985) The 1985 catalog of mapped genes and report of the nomenclature committee. Cytogenet Cell Genet 40:8–66

McAlpine PJ, Boucheix C, Pakstis AJ, Stranc LC, Berent TG, Shows TB (1988) The 1988 catalog of mapped genes and report of the nomenclature committee. Cytogenet Cell Genet 49:4–38

McElreavey KD, Vilain E, Boucekkine C, Vidaud M, Jaubert F, Richaud F, Fellous M (1992) XY sex reversal associated with a nonsense mutation in *SRY*. Genomics 13: 838–840

McKay RDG, Heritage J, Bobrow M, Cooke HJ (1978) Endonuclease analysis of Y chromosome DNA. Cytogenet Cell Genet 22:357–358

McLaren A (1985) Relation of germ cell sex to gonadal differentiation. In: Halvorson HO, Monroy A (eds) The origin and evolution of sex, vol 7. Liss, New York, pp 289–300

McLaren A, Ferguson-Smith MA (eds) (1988) Sex determination in mouse and man. Philos Trans R Soc Lond [Biol] 322:1–157

McLaren A, Simpson E, Tomonari K, Chandler P, Hogg H (1984) Male sexual differentiation in mice lacking H-Y antigen. Nature 312:552–555

Migeon BR, Shapiro LJ, Norum RA, Mohandas T, Axelman J, Dabora RL (1982) Differential expression of steroid sulphatase locus on active and inactive human X chromosome. Nature 299: 838–840

Mitchell MJ, Woods DR, Tucker PK, Opp JS, Bishop CE (1991) Homology of a candidate spermatogenic gene from the mouse Y chromosome to the ubiquitin-activating enzyme E1. Nature 354: 483–486

Mittwoch U (1992) Sex determination and sex reversal: genotype, phenotype, dogma and semantics. Hum Genet 89:467–479

Morton NE (1991) Parameters of the human genome. Proc Natl Acad Sci USA 88: 7474–7476

Müller G, Schempp W (1989) Mapping the human ZFX locus to Xp21.3 by in situ hybridization. Hum Genet 82:82–84

Müller G, Schempp W (1991) Comparative mapping of ZFY in the hominoid apes. Hum Genet 88: 59–63

Müller U (1987) Mapping of testis-determining locus on Yp by the molecular genetic analysis of XX males and XY females. Development 101 [Suppl]: 51–58

Müller U, Urban E (1981) Reaggregation of rat gonadal cells in vitro: experiments on the function of H-Y antigen. Cytogenet Cell Genet 31:104–107

Müller U, Donlon T, Schmid M, Fitch N, Richer C-L, Lalande M, Latt SA (1986a) Deletion mapping of the testis determining locus with DNA probes in 46,XX males and in 46,XY and 46,X,dic(Y) females. Nucleic Acids Res 14:6489–6505

Müller U, Lalande M, Donlon T, Latt SA (1986b) Moderately repeated DNA sequences specific for the short arm of the human Y chromosome are present in XX males and reduced in copy number an XY female. Nucleic Acids Res 14:1325–1340

Müller U, Latt SA, Donlon T (1987) Y-specific DNA sequences in male patients with 46,XX and 47,XXX karyotypes. Am J Med Genet 28:393–401

Müller U, Wachtel SS (1991) Are testis-secreted H-Y serological antigen and Müllerian inhibiting substance the same? Am J Hum Genet 49:S 22

Nagamine CM, Chan K, Kozak CA, Lau Y-F (1989) Chromosome mapping and expression of a putative testis-determining gene in mouse. Science 243:80–83

Nakagome Y, Seki S, Fukutani K, Nagafuchi S, Nakahori Y, Tamura T (1991) PCR detection of distal Yp sequences in an XX true hermaphrodite. Am J Med Genet 41:112–114

Nakahori Y, Nakagome Y (1989) Conserved Xp-Yp homologous region which encodes open reading frames corresponding to "amelogenin". Cytogenet Cell Genet 51:1050

Nakahori Y, Mitani K, Yamada M, Nakagome Y (1986) A human Y-chromosome-specific repeated DNA family (DYZ1) consists of a tandem array of pentanucleotides. Nucleic Acids Res 14:7569–7580

Nakahori Y, Takenaka O, Nakagome Y (1991a) A human X-Y homologous region encodes "amelogenin". Genomics 9:264–269

Nakahori Y, Tamura T, Nagafuchi S, Fujieda K, Minowada et al. (1991b) Molecular cloning and mapping of 10 new probes on the human Y chromosome. Genomics 9:765–769

Nasrin N, Buggs C, Kong XF, Carnazza J, Goebl M, Alexander-Bridges M (1991) DNA-binding properties of the product of the testis determining gene and a related protein. Nature 354:317–320

Nomiyama H, Obaru K, Jinno Y, Matsuda I, Shimada K, Miyata T (1986) Amplification of human argininosuccinate synthetase pseudogenes. J Mol Biol 192:221–233

Oakey R, Tyler-Smith C (1990) Y chromosome DNA haplotyping suggests that most European and Asian men are descended from one of two males. Genomics 7:325–330

O'Brien SJ, Graves JAM (1991) Report of the committee on comparative gene mapping. Cytogenet Cell Genet 58:1124–1151

Ohno S (1967) Sex chromosomes and sex-linked genes. Springer, Berlin Heidelberg New York

Ohno S (1976) Major regulatory genes for mammalian sexual development. Cell 7:315–321

Ohno S (1979) Major sex-determining genes. Springer, Berlin Heidelberg New York

Ohno S, Beçak W, Beçak ML (1964) X-autosome ratio and the behavior pattern of individual X chromosomes in placental mammals. Chromosoma 15:14–30

Olmo E, Cobror O, Morescalchi A, Odierna G (1984) Homomorphic sex chromosomes in the lacertid lizard *Takydromus sexlineatus*. Heredity (Edinburgh) 53:457–459

Ostrer H, Wright G, Clayton CM, Skordis NA, MacGillivray MH (1989) Familial XX chromosomal maleness does not arise from a Y-chromosomal translocation. J Pediatr 114:977–982

Page DC (1987) Hypothesis: a Y-chromosomal gene causes gonadoblastoma in dysgenetic gonads. Development 101 [Suppl]:151–155

Page DC (1988) Is *ZFY* the sex-determining gene on the human Y chromosome? Philos Trans R Soc Lond [Biol] 322:155–157

Page DC, Martinville B de, Barker D, Wyman A, White R, Francke U, Botstein D (1982) Single-copy sequence hybridizes to polymorphic and homologous loci on human X and Y chromosomes. Proc Natl Acad Sci USA 79:5352–5356

Page DC, Harper ME, Love J, Botstein D (1984) Occurrence of a transposition from the X-chromosome long arm to the Y-chromosome short arm during human evolution. Nature 311:119–123

Page DC, de la Chapelle A, Weissenbach J (1985) Chromosome Y-specific DNA in related human XX males. Nature 315:224–226

Page DC, Bieker K, Brown LG, Hinton S, Leppert M, Lalouel J-M, Lathrop M, Nystrom-Lahti M, de la Chapelle A, White R (1987a) Linkage, physical mapping, and DNA sequence analysis of pseudoautosomal loci on the human X and Y chromosomes. Genomics 1:243–256

Page DC, Brown LG, de la Chapelle A (1987b) Exchange of terminal portions of X- and Y-chromosomal short arms in human XX males. Nature 328:437–440

Page DC, Mosher R, Simpson EM, Fisher EMC, Mardon G, Pollak J, McGillivray B, de la Chapelle A, Brown LG (1987c) The sex determining region of the human Y chromosome encodes a finger protein. Cell 51:1091–1104

Page DC, Disteche CM, Simpson EM, de la Chapelle A, Andersson M, Alitalo T, Brown LG, Green P, Akots G (1990a) Chromosomal localization of *ZFX* – a human gene that escapes X inactivation – and its murine homologs. Genomics 7:37–46

Page DC, Fisher EMC, McGillivray B, Brown LG (1990b) Additional deletion in sex-determining region of human Y chromosome resolves paradox of X,t(Y;22) female. Nature 346:279–281

Palmer MS, Sinclair AH, Berta P, Ellis NA, Goodfellow PN, Abbas NE, Fellous M (1989) Genetic evidence that *ZFY* is not the testis-determining factor. Nature 342:937–939

Palmer MS, Berta P, Sinclair AH, Pym B, Goodfellow PN (1990) Comparison of human *ZFY* and *ZFX* transcripts. Proc Natl Acad Sci USA 87:1681–1685

Pedicini A, Camerino G, Avarello R, Guioli S, Zuffardi O (1991) Probe St35-239 (DXYS64) reveals homology between the distal ends of Xq and Yq. Genomics 11: 482–483

Petit C, de la Chapelle A, Levilliers J, Castillo S, Noël B, Weissenbach J (1987) An abnormal terminal X-Y interchange accounts for most but not all cases of human XX maleness. Cell 49:595–602

Petit C, Levilliers J, Weissenbach J (1988) Physical mapping of the human pseudo-autosomal region; comparison with genetic linkage map. EMBO J 7:2369–2376

Petit C, Levilliers J, Rouyer F, Simmler MC, Herouin E, Weissenbach J (1990) Isolation of sequences from Xp22.3 and deletion mapping using sex chromosome rearrangements from human X-Y interchange sex reversals. Genomics 6:651–658

Ramsay M, Bernstein R, Zwane E, Page DC,Jenkins T (1988) XX true hermaphroditism in Southern African blacks: an enigma of primary sexual differentiation. Am J Hum Genet 43:4–13

Rappold GA, Lehrach H (1988) A long-range restriction map of the pseudoautosomal region by partial digest PFGE analysis from the telomere. Nucleic Acids Res 12: 5361–5377

Rappold GA, Henke A, Klink A, Horsthemke B, Wapenaar M, Armour J, Gough N (1991) Patients with deletions within the pseudoautosomal region help map 13 new pseudoautosomal probes. Cytogenet Cell Genet 58:2083

Ray-Chaudhuri SP, Singh L, Sharma T (1970) Sexual dimorphism in somatic interphase nuclei of snakes. Cytogenetics 9:410–423

Ray-Chaudhuri SP, Singh L, Sharma T (1971) Evolution of sex-chromosomes and formation of W-chromatin in snakes. Chromosoma 33:239–251

Roberts L (1988) Zeroing in on the sex switch. Science 239:21–23

Rofe R, Hayman D (1985) G-banding evidence for a conserved complement in the Marsupialia. Cytogenet Cell Genet 39:40–50

Rosenfeld RG, Luzzati L, Hintz RL, Miller OJ, Koo GC, Wachtel SS (1979) Sexual and somatic determinants of the human Y chromosome: studies in a 46,XYp- phenotypic female. Am J Hum Genet 31:458–468

Rouyer F, Simmler M-C, Johnsson C, Vergnaud G, Cooke HJ, Weissenbach J (1986) A gradient of sex linkage in the pseudoautosomal region of the human sex chromosomes. Nature 319:291–295

Rouyer F, Simmler M-C, Page DC, Weissenbach J (1987) A sex chromosome rearrangement in a human XX male caused by Alu-Alu recombination. Cell 51:417–425

Rouyer F, de la Chapelle A, Andersson M, Weissenbach J (1990) An interspersed repeated sequence specific for human subtelomeric regions. EMBO J 9:505–514

Salido EC, Yen PH, Koprivnikar K, Yu L-C, Shapiro LJ (1992) The human enamel protein gene amelogenin is expressed from both the X and Y chromosomes. Am J Hum Genet 50:303–316

Sandberg AA (ed) (1985) The Y chromosome, parts A, B. Liss, New York

Schempp W, Meer B (1983) Cytologic evidence for three human X-chromosomal segments escaping inactivation. Hum Genet 63:171–174

Schempp W, Schmid M (1981) Chromosome banding in Amphibia VI. Chromosoma 83: 697–710

Schempp W, Toder R (1992) Molecular cytogenetic studies on the evolution of sex chromosomes in primates. In: Graves JAM, Reed K (eds) Proceedings of the 1992 Boden research conference: mammalian sex chromosomes and sex-determining genes: their differentiation, autonomy and interactions in gonad differentiation and function. Harwoord, New York, (in press)

Schempp W, Weber B, Serra A, Negri G, Gal A, Wolf U (1985) A 45,X male with evidence of a translocation of Y euchromatin onto chromosome 15. Hum Genet 71:150–154

Schempp W, Müller G, Scherer G, Bohlander SK, Rommerskirch W, Fraccaro M, Wolf U (1989a) Localization of Y chromosome sequences and X chromosomal replication studies in XX males. Hum Genet 81:144–148

Schempp W, Weber B, Müller G (1989b) Mammalian sex-chromosome evolution: a conserved homologous segment on the X and Y chromosomes in primates. Cytogenet Cell Genet 50: 201–205

Scherer G, Schempp W, Baccichetti C, Lenzini E, Dagna Bricarelli F, Doria Lamba Carbone L, Wolf U (1989a) Duplication of an Xp segment that includes the ZFX locus causes sex inversion in man. Hum Genet 81:291–294

Scherer G, Schempp W, Fraccaro M, Bausch E, Bigozzi V, Maraschio P, Montali E, Simoni G, Wolf U (1989b) Analysis of two 47,XXX males reveals X-Y interchange and maternal or paternal nondisjunction. Hum Genet 81:247–251

Schmeckpeper BJ, Willard HF, Smith KD (1981) Isolation and characterization of cloned human DNA fragments carrying reiterated sequences common to both autosomes and the X chromosome. Nucleic Acids Res 8:1853–1872

Schmid M, Guttenbach M, Nanda I, Studer R, Epplen JT (1990) Organization of DYZ2 repetitive DNA on the human Y chromosome. Genomics 6:212–218

Schneider-Gädicke A, Beer-Romero P, Brown LG, Mardon G, Luoh S-W, Page DC (1989a) Putative transcription activator with alternative isoforms encoded by human ZFX gene. Nature 342:708–711

Schneider-Gädicke A, Beer-Romero P, Brown LG, Nussbaum R, Page DC (1989b) ZFX has a gene structure similar to ZFY, the putative human sex determinant, und escapes X inactivation. Cell 57:1247–1258

Schwarzacher-Robinson T, Cram LS, Meyne J, Moyzis RK (1988) Characterization of human heterochromatin by in situ hybridization with satellite DNA clones. Cytogenet Cell Genet 47:192–196

Shapiro LJ (1985) Steroid sulfatase deficiency and the genetics of the short arm of the human X chromosome. In: Harris H, Hirschhorn K (eds) Advances in human genetics. Plenum, New York, pp 331–381

Shapiro LJ, Mohandas T, Weiss R, Romeo G (1979) Non-inactivation of an X-chromosome locus in man. Science 204:1224–1226

Shows TB, McAlpine PJ (1982) The 1981 catalogue of assigned human genetic markers and report of the nomenclature committee. Cytogenet Cell Genet 32:221–245

Silver J, Rabson A, Bryan T, Willey R, Martin MA (1987) Human retroviral sequences on the Y chromosome. Mol Cell Biol 7:1559–1562

Silvers WK, Gasser DL, Eicher EM (1982) H-Y antigen, serologically detectable male antigen and sex determination. Cell 28:439–440

Simmler M-C, Rouyer F, Vergnaud G, Nyström-Lahti M, Ngo KY, de la Chapelle A, Weissenbach J (1985) Pseudoautosomal DNA sequences in the pairing region of the human sex chromosomes. Nature 317:692–697

Simpson E, Chandler P, Goulmy E, Disteche CM, Ferguson-Smith MA, Page DC (1987) Separation of the genetic loci for the H-Y antigen and for testis determination on human Y chromosome. Nature 326:876–878

Simpson JL (1976) Disorders of sexual differentiation. Etiology and clinical delineation. Academic Press, New York

Simpson JL, Blagowidow N, Martin AO (1981) XY gonadal dysgenesis: genetic heterogeneity based upon clinical observations, H-Y antigen status and segregation analysis. Hum Genet 58:91–97

Sinclair AH, Foster JW, Spencer JA, Page DC, Palmer M, Goodfellow PN, Graves JAM (1988) Sequences homologous to ZFY, a candidate human sex-determining gene, are autosomal in marsupials. Nature 336:780–783

Sinclair AH, Berta P, Palmer MS, Hawkins JR, Griffiths BL, Smith MJ, Foster JW, Frischauf A-M, Lovell-Badge R, Goodfellow PN (1990) A gene from the human sex-determining region encodes a protein with homology to a conserved DNA-binding motif. Nature 346:240–244

Singh L, Purdom IF, Jones KW (1976) Satellite DNA and evolution of sex chromosomes. Chromosoma 59:43–62

Singh L, Purdom IF, Jones KW(1979) Behaviour of sex chromosome-associated satellite DNAs in somatic and germ cells in snakes. Chromosoma 71:167–181

Singh L, Purdom IF, Jones KW (1980) Sex chromosome-associated satellite DNA: evolution and conversation. Chromosoma 79:137–157

Skordis NA, Stetka DG, McGillivray MH, Greenfield SP (1987) Familial 46,XX males coexisting with familial 46,XX true hermaphrodites in same pedigree. J Pediatr 100: 244–248

Smith KD, Young KE, Talbot CC Jr, Schmeckpeper BJ (1987) Repeated DNA of the human Y chromosome. Development 100 [Suppl]:77–92

Solari AJ (1980) Synaptonemal complexes and associated structures in microspread human spermatocytes. Chromosoma 81:315–337

Soriano P, Keitges EA, Schorderet DF, Harbers K, Gartler SM, Jaenisch R (1987) High rate of recombination and double crossovers in the mouse pseudoautosomal region during male meiosis. Proc Natl Acad Sci USA 84:7218–7220

Speed RM, Chandley AC (1990) Prophase of meiosis in human spermatocytes analysed by EM microspreading in infertile men and their controls and comparisons with human oocytes. Hum Genet 84:547–554

Spencer JA, Sinclair AH, Watson JM, Graves JAM (1991) Genes on the short arm of the human X chromosome are not shared with the marsupial X. Genomics 11:339–345

Su H, Kozak CA, Veerhuis R, Lau Y-FC, Wiberg U (1992) Isolation of a phylogenetically conserved and testis-specific gene using a monoclonal antibody against the serological H-Y antigen. J Reprod Immunol 21:275–291

Su T-S, Nussbaum RL, Airhart S, Ledbetter DH, Mohandas T, O'Brien WE, Beaudet AL (1984) Human chromosome assignments for 14 argininosuccinate synthetase pseudogenes: cloned DNAs as reagents for cytogenetic analysis. Am J Hum Genet 36:954–964

Tiepolo L, Zuffardi O (1976) Localization of factors controlling spermatogenesis in the nonfluorescent portion of the human Y chromosome long arm. Hum Genet 34:119–124

Tyler-Smith C, Brown WRA (1987) Structure of the major block of alphoid satellite DNA on the human Y chromosome. J Mol Biol 195:457–470

van de Wetering M, Oosterwegel M, Dooijes D, Clevers H (1991) Identification and cloning of TCF-1, a T-lymphocyte specific transcription factor containing a sequence-specific HMG box. EMBO J 10:123–132

van Niekerk WA, Retief AE (1981) The gonads of human true hermaphrodites. Hum Genet 58:117–122

Vergnaud G, Page DC, Simmler M-C, Brown L, Rouyer F, Noel B, Botstein D, de la Chapelle A, Weissenbach J (1986) A deletion map of the human Y chromosome based on DNA hybridization. Am J Hum Genet 38:109–124

Vigier B, Forest MG, Eychenne B, Bézard J, Garrigou O, Robel P, Josso N (1989) Anti-Müllerian hormone produces endocrine sex reversal of fetal ovaries. Proc Natl Acad Sci USA 86:3684–3688

Vilain E, McElreavey KD, Jaubert F, Raymond J-P, Richaud F, Fellous M (1992) Familial case with sequence variant in the testis-determining region associated with two sex phenotypes. Am J Hum Genet 50:1008–1011

Vogt P, Henning W (1986) Molecular structure of the lampbrush loop nooses of the Y chromosome of *Drosophila hydei:* I. The Y chromosome-specific repetitive DNA sequence family ay1 is dispersed in the loop DNA. Chromosoma 94:449–458

Vogt P, Keil R, Köhler M, Lengauer C, Lewe D, Lewe G (1991) Seletion of DNA sequences from interval 6 of the human Y chromosome with homology to a Y chromosomal fertility gene sequence of *Drosophila hydei.* Hum Genet 86:341–349

Vogt P, Chandley AC, Hargreave TB, Keil R, Ma K, Sharkey A (1992) Microdeletions in intervals 6 of the Y chromosome of males with idiopathic sterility point to disruption of AZF, a human spermatogenesis gene. Hum Genet 89:491–496

Wachtel SS (1983) H-Y antigen and the biology of sex determination. Grune and Stratton, New York

Wachtel SS (ed) (1989) Evolutionary mechanisms in sex determination. CRC Press, Boca Raton

Wachtel SS, Ohno S, Koo GC, Boyse EA (1975) Possible role for H-Y antigen in the primary determination of sex. Nature 257:235–236

Waibel F, Scherer G, Fraccaro M, Hustinx TWJ, Weissenbach J, Wieland J, Mayerová A, Back E, Wolf U (1987) Absence of Y-specific DNA sequences in human 46,XX true hermaphrodites and in 45,X mixed gonadal dysgenesis. Hum Genet 76:332–336

Walter MA, Gubbay J, Capel B, Lovell-Badge R, Goodfellow P (1991) Sry is the testis-determining factor: evidence from the expression of the Sry gene in COS cells. Am J Hum Genet 49:S422

Watson JM (1990) Monotreme genetics and cytology and a model for sex chromosome evolution. Aust J Zool 37:385–406

Watson JM, Spencer JA, Riggs AD, Graves JAM (1990) The X chromosome of monotremes shares a highly conserved region with the eutherian and marsupial X chromosomes despite the absence of X chromosome inactivation. Proc Natl Acad Sci USA 87: 7125–7129

Watson JM, Spencer JA, Riggs AD, Graves JAM (1991) Sex chromosome evolution: platypus gene mapping suggests that part of the human X chromosome was originally autosomal. Proc Natl Acad Sci USA 88:11256–11260

Weber B, Schempp W, Wiesner H (1986) An evolutionarily conserved early-replicating segment on the sex chromosomes of man and the great apes. Cytogenet Cell Genet 43: 72–78

Weber B, Weissenbach J, Schempp W (1987) Conservation of human-derived pseudoautosomal sequences on the sex chromosomes of the great apes. Cytogenet Cell Genet 45: 26–29

Weissenbach J (1988) Mapping the human Y chromosome. Philos Trans R Soc Lond [Biol] 322:125–131

Weissenbach J, Goodfellow PN (1991) Report of the committee on the genetic constitution of the Y chromosome. Cytogenet Cell Genet 58:967–985

Weissenbach J, Levilliers J, Petit C, Rouyer F, Simmler M-C (1987) Normal and abnormal interchanges between the human X and Y chromosomes. Development 101 [Suppl]: 67–74

Weissenbach J, Goodfeloow PN, Smith KD (1989) Report of the commitee on the genetic constitution of the Y chromosome. Cytogenet Cell Genet 51:438–449

Wevrick R, Willard HF (1989) Long-range organization of tandem arrays of α satellite DNA at the centromeres of human chromosomes: high-frequency array-length polymorphism and meiotic stability. Proc Natl Acad Sci USA 86:9394–9398

Wiberg UH (1987) Facts and considerations about sex-specific antigens. Hum Genet 76: 207–219

Wolf U (1978) Zum Mechanismus der Gonadendifferenzierung. Bull Schweiz Akad Med Wiss 34:357–368

Wolf U (1981) Genetics of primary gonadal differentiation. In: Frajese G et al. (eds) Oligozoospermia: recent progress in andrology. Raven, New York, pp 225–231

Wolf U (1985) Genes of the H-Y antigen system and their expression in mammals. In: Sandberg AA (ed) The Y chromosome: A. Basic characteristics of the Y chromosome. Liss, New York, pp 81–91

Wolf U (1988) Sex inversion as a model for the study of sex determination in vertebrates. Philos Trans R Soc Lond [Biol] 322:97–107

Wolfe J, Erickson RP, Rigby PWJ, Goodfellow PN (1984) Cosmid clones derived from both euchromatic and heterochromatic regions of the human Y chromosome. EMBO J 3: 1997–2003

Wolfe J, Darling SM, Erickson RP, Craig RW, Buckle VJ, Rigby PWJ, Willard HF, Goodfellow PN (1985) Isolation and characterization of an alphoid centromeric repeat family from the human Y chromosome. J Mol Biol 182:477–485

Wrigley JM, Graves JAM (1988) Sex chromosome homology and incomplete, tissue-specific X-inactivation suggest that monotremes represent an intermediate stage of mammalian sex chromosome evolution. J Hered 79:115–118

Yamada K, Isurigi K (1981) H-Y antigen studies in thirty patients with abnormal gonadal differentiation: correlations among sex chromosome complement, H-Y antigen, and gonadal type. Jpn J Hum Genet 26:227–235

Yen PH, Allen E, Marsh B, Mohandas T, Wang N, Taggart RT, Shapiro LJ (1987) Cloning and expression of steroid sulfatase cDNA and the frequent occurrence of deletions in STS deficiency: implications for X-Y interchange. Cell 49:443–454

Yen PH, Marsh B, Allen E, Tsai S-P, Ellison J, Connolly L, Neiswanger K, Shapiro LJ (1988) The human X-linked steroid sulfatase gene and a Y-encoded pseudogene: evidence for an inversion of the Y chromosome during primate evolution. Cell 55: 1123–1135

Yen PH, Tsai S-P, Wanger SL, Steele MW, Mohandas TK, Shapiro LJ (1991) X/Y translocations resulting from recombination between homologous sequences on Xp and Yq. Proc Natl Acad Sci USA 88:8944–8948

Zech L (1969(Investigation of metaphase chromosomes with DNA-binding fluorochromes. Exp Cell Res 58:463

Zenzes MT, Müller U, Aschmoneit I, Wolf U (1978a) Studies on H-Y antigen in different cell fractions of the testis during pubescence. Hum Genet 45:297–303

Zenzes MT, Wolf U, Engel W (1978b) Organization in vitro of ovarian cells into testicular structures. Hum Genet 44:333–338

Subject Index